HVAC GUIDE

FOR BEGINNERS

2023 EDITION

INTRODUCTION

HVAC systems are amazing devices that can extract heat energy from something seemingly cold - like the air or the ground - and use it for heating. Or it can extract heat from something hot (like air) very fast and make it cold. This concept is somewhat confusing and goes against our natural understanding of how things should work. Like any other technology, HVAC systems can perform poorly if installed improperly or used incorrectly, and in this case, you can't expect such system to deliver what the manufacturer promises.

The world's energy demand for oil, gas and coal is steadily increasing in line with population growth and the energy requirements of the growing global economy. Unfortunately, the global energy transition is just beginning and is far from complete, especially in developing countries.

This book is intended for anyone who wants to learn or better understand the subject of heat pumps, such as beginners, homeowners, students or someone who consider starting a job as HVAC contractor and wants to get understanding of what HVAC is and how it works.

When installing a HVAC system, small details can make the difference between excellent and mediocre performance. To get a handle on these small details that make all the difference, you need to understand at least a whole host of different ideas, some of which may be new to you. Once you understand the basic concepts, you can put them together like a puzzle.

What is HVAC?

Heating, Ventilation, and Air Conditioning (HVAC) – in simple terms, it is a technology whose main goal is to create a comfortable environment for people including proper temperature, ventilation, and air quality in the building. Good ventilation decreases moisture levels in the air and lowers the risk of mold, bacterias, allergens, and other problems.

An air conditioner or heat pump is a device that can transfer heat from one place to another. The main advantage of a heat pump is that it can heat and cool a house. The physical effect used in heat pumps was discovered by William Cullen in 1777. And the first refrigeration machine based on this effect was developed just 60 years later.

The heat supplied by a heat pump is not free, as a certain amount of energy is required to operate the system. Interestingly, however, heat pumps can deliver much more usable heat energy than is required to operate them. Until now, fuel has been cheap and the impact on the environment has not yet been considered serious or important.

History of Heat Pump / Refrigerator

The first refrigerator was built by William Cullen. The scientist presented it at the University of Glasgow in 1748. However, his invention was not successful.

In 1818, Jacob Perkins was granted a patent for the "first refrigerating circuit." Jacob Perkins' idea was based on the same idea as William Cullen's. The refrigerator did not become popular for home use until 150 years later.

In 1852, William Thompson (Lord Kelvin), propagated the idea of a heat pump. His concept was based on an "inverted heat engine". Based on Lord Kelvin's idea, Peter von Rittinger developed the first heat pump in 1856. Then the studies in this field were stopped because of World War II and lack of resources.

In 1945, the Norwegian electrician John Sumner built a radical new heat pump system. This system was successful but failed to catch on due to low prices for alternative energy sources (oil and gas).

Energy Demand and Energy Basics

In most modern houses with good insulation, about 50% of the required heat energy is used for water heating and the remaining 50% for space heating. In old houses without high-quality insulation, the cost of heating the rooms can be 60-70% of the cost of the required thermal energy.

In addition, poor insulation also has a bad effect on the cooling of the house - the walls and ceiling of the house quickly receive heat from the outside, and the temperature in the house starts to rise very fast. Your HVAC system must work hard to keep the building at the right temperature. The hotter outside the harder the system should work.

As a result, this will lead to high electricity costs, and if we look at a period of 20 years, this will lead to very large amounts of money being paid to utilities.

Now let's look on some basic physic that we should know.

Electric current - amperage [I] (amps) - the amount of electricity that "flows" in a cable/wire. To conduct a lot of current (amperes), **we need** thick wires with a large cross-section.

The electrical voltage [U] - the potential (volts) - is the force with which electrons flow through the wire. The electrical potential is always present in a circuit - so the voltage is always present even when no current is flowing.

Electrical power [P] (watts) - the rate at which electrical energy is transferred in a circuit per unit time.

P = I x U

P = Amps x Volts [Watts]

Example: The electrical power at a voltage of 2 volts and a current of 3 amperes is 6 watts. P = 2 V x 3 A = 6 W

1000 Watt [W] = 1 Kilowatt [kW]

Energy, measured in watt-hours [Wh], is the product of power [W] and time, measured in hours [h].

Energy = P x t = Wh

Energy = Watt x Hour = [Wh]

Efficiency [N]

Efficiency - the percentage of energy that a particular mechanism can use from the energy source to perform a task. The greater the amount of energy that can be used, the higher the efficiency.

N = usable energy / total energy

N = 100 kWh / 100 kWh = 1;

Efficiency 1 = 100 %;

Efficiency 0.8 = 80 %;

Space heating demand - the amount of heat energy required to maintain the desired temperature in the space.

Heating energy demand is measured per square meter and per year of the residential building.

Example:

In a well-insulated new house, the demand for thermal energy is about 50 kWh per square meter per year.

If the area of our house is 100 square meters:

$100\ m^2 \times 50\ kWh/m^2$ per year = 5000 kWh per year

For example, we heat our house with oil. And 1 liter of oil contains 9.8 kWh of energy.

Then per year:

5000 kWh / 9.8 kWh = 510 liters (oil)

Here it is assumed that 100% of the energy is converted into heat and used for heating!

In reality, the usable energy in oil heating method is about 80% (**efficiency - 0.8**).

5000 kWh / 0.8 = 6250 kWh

6250 kWh / 9.8 kWh = 638 liters (oil)

For example, a photovoltaic panel has an efficiency of about 20%, which means that only 20% of solar energy is converted into electricity. If the panel converts 100% of the solar energy into electricity (100% efficiency), then a 100 W panel would generate 500 W of power.

What is pressure? Pressure is the force acting on a unit area. Pressure is measured in pascals **[Pa]**. It is indicated by the letter **p**.

Pressure = Force divided by Area:

[Pressure] = F [Force] / A [Area] = p [Pascals].

$p = F / A$

$1\ Pa = 1\ N / 1\ m^2$

100 000 Pa = 1 bar

Atmospheric pressure:

1 atm = 1.0133 bar

1 bar and 1 atmosphere is almost equal.

A pressure that is less than 1 atmosphere is called a vacuum **(negative pressure).**

A pressure of more than 1 atmosphere is called overpressure or just pressure.

You can also think of gas pressure as the number of particles contained in a vessel of a certain volume. The more particles (gas molecules), the greater the pressure. The molecules are also constantly in motion and collide with the walls of the vessel. They exert pressure on the walls of the vessel. The pressure is evenly distributed in all directions there.

In simplified terms, this can be illustrated as follows:

Dependence of the boiling point of a liquid on pressure

When the pressure drops, the boiling point of liquid also drops:

1 atm = 100 °C

0.5 atm = 80.8 °C

0.01 atm = 6.69 °C

When the gas expands, the gas temperature of that gas decreases.

The boiling point increases with increasing pressure under that liquid:

2 atm = 119.6 °C

5 atm = 151.1 °C

10 atm = 179 °C

When gas is strongly compressed, its temperature increases.

BTU [British Thermal Unit] - is a measure of the heat content of fuels or energy sources. 1 BTU is the quantity of heat required to raise the temperature of one pound of liquid water by 1° Fahrenheit (F) at the temperature that water has its greatest density (approximately 39° F). This definition is given by *U.S. Energy Information Administration.*

Mechanisms Of Heat Transfer

1. **Convection** –the movement of liquids or gases due to temperature differences. For example, hot air from home radiators always moves upward in a room. This is because heated air has a lower air mass than cold air.
2. **Radiation -** heat is exchanged by means of electromagnetic waves.
3. **Phase change** – heat transfer occurs when a substance changes its phase state. By using these mechanisms work all air conditioners. Refrigerant change its state from liquid to gas and in such a way absorbs the heat because become cold.
4. **Conduction** –the transfer of heat between two objects that are in direct contact with each other. Conductivity depends on the thermal conductivity of the material of the objects with which they come into contact.

Now that you've reminded yourself of some basic physics knowledge from school, let's move on.

CHAPTER 1: HEAT PUMPS / REFRIGERATION - THE BEGINNING

HVAC Types

1. **Split systems.** This system has two units: outdoor and indoor. Also, on the market are available mini-split systems (ductless). In this case, there is only one outdoor unit which is connected to many indoor units.
2. **Central systems.** This system is commonly used in large buildings.
3. **Air and geothermal heat pump systems.** Heat Pumps are essentially AC but work in another direction. They gather heat outside from the air or ground and pump it inside the building.
4. **Packaged unit system.** Typically, is used when there is limited space inside the building. All components of the system are installed in the outdoor unit of the building.
5. **Variable Refrigerant Flow Systems.** This system allows to have multiple indoor units connected to one outdoor unit. Allow heating or cooling for different parts of the building simultaneously.

If you must choose between a heat pump and a conventional heating system or if you must choose between installing simple air conditioning (AC) system and a heat pump - you should go for a heat pump. Heat pumps also have the advantage of being versatile: You can use them for both heating in winter and for cooling in the summer. This means you don't need two separate systems in the building.

Heat pump efficiency?

Most heat pumps operate less efficiently when the outdoor temperature drops below 50 °F (10 degrees Celsius). This is true for air-source heat pumps, whose efficiency is highly dependent on the outdoor temperature.

For other types of heat pumps, where the heat source is a more stable medium such as groundwater or soil, the efficiency is much less dependent on the ambient temperature.

Where do I start if I want to install a heat pump for heating my house?

Everyone strives for the highest energy efficiency with the least effort and installation costs.

There are a few basic things you need to do before you even think about installing a heat pump or any other HVAC system.

1. Determine your heating needs (this is usually done by the installation company or the energy consultant).
2. Check the heating system already installed in the building. It may need to be dismantled or modified to accommodate the heat pump system. The installer should be informed of any additional preparations that need to be made.
3. Contact a certified heat pump company to decide what type and size of the system is needed.
4. Find out about the government subsidies currently available in your area.
5. Improve your home's insulation to reduce your heat and cold losses and therefore heating/cooling demand in your building. Consider doing insulation or additional insulation to the building. Insulation also reduces your energy bills during the year.

After doing all this, we can now compare prices for your heat pump system with different installation companies and choose the best option for you.

Reducing energy demand is always the best place to start. Your overall energy strategy should start by trying to minimize your total energy demand and heat loss. Insulation is always a good option, even if it is very expensive. Over time, it will pay for itself.

Use the thermal imaging camera to determine where your building is losing heat or accept it from outside during the summer. You can buy or borrow it. This will give you an idea of your home's energy efficiency and what needs to be improved. The best time to have this done is during the winter. However, you can also try to do it on cold days or nights when the temperature difference between the indoor and outdoor temperature of the building is greater. The difference of 20 degrees should show you where the heat or cold goes.

Consider insulating your roof. Because the roof is the place where most heat is lost, about 30-40% of all heat/cold loss. Heat rises and accumulates first in the upper part of the room under the ceiling. The room begins to warm up from the top down.

Consider insulating the walls of the building and replacing old windows with new, more energy-efficient ones. The walls are in second place with the most heat loss - about 20-30%.

If the difference between the window insulation and the wall insulation is very large, the warm air can condense in the cold walls. This can lead to mold.

If we improve the insulation of a house, we need less heating/cooling load for the building and we can lower the temperature of the heating or cooling, which leads to higher efficiency of the HVAC system. More details about this will be discussed in the following chapters.

A practical understanding of heat

By definition, heat is the kinetic energy of atoms, molecules, or other particles that make up an object. Thermal energy is transferred from one object to another.

How can heat be extracted from something that is cold?

Heat pumps work on the principle of refrigeration machines (e.g. air conditioners, refrigerators, etc.). The principle of operation of the heat pump is based on the **second law of thermodynamics**.

This law states that heat moves when there is a temperature difference between two objects in contact - from a higher-temperature object to a lower-temperature object until the temperature at the two places (objects) become equal.

+10 >>> +5

After that, the temperature of the two objects become the same:

+ 7 === +7

Is 0 degrees Celsius really 0?

We are used to thinking that 0 degrees Celsius (32 °F) means the absence of heat, but in reality, there is a lot of heat at 32 °F, it is simply the temperature at which water freezes.

The temperature at which the amount of heat is 0 is -273.15 degrees Celsius or minus 459.76 Fahrenheit. In most places in space, the temperature is exactly -273.15 degrees Celsius. This temperature is also called absolute zero. For this reason, the Kelvin scale is often used in physics, where 0 degrees Celsius equals 273.15 K.

0 degrees Kelvin = -273.15 degrees Celsius or -459.76 Fahrenheit.

Once we understand this, we know that when the outside temperature is at 32 °F (0 degrees Celsius), there is still 273 degrees Celsius or almost 250 °F of heat available. Not bad, is it?

It is therefore possible to extract heat from an environment below 0 degrees Celsius. We just need to use the laws of nature and technical progress to our advantage.

There is a surplus of thermal energy.

Let's take a look at the thermal energy supplied by the sun.

The maximum energy absorbed from the sun is 1000 watts (1 kW) per square meter of the surface directly exposed to the sun.

The energy comes from the sun's rays falling on the ground and being converted into heat energy. The heated surface then begins to conduct the heat into the interior of the soil, where it accumulates.

The amount of "geothermal" heat rising from the earth's core is tiny; in most places, it is only 1/20 of a watt per square meter (or about 10 ft^2). This is very small compared to the radiation from the sun, which is about 1000 W/m^2.

The Heat Energy Sources for HVAC Systems.

Air and Water

The temperature of outdoor air generally varies from -4 °F (-20°C) in winter to +86 °F (+30°C) in summer, depending on where you live. This does not make it an ideal heat source, as the air is coldest when the demand for heat is greatest. Still, it can be a viable and effective heat source when the outside temperature doesn't drop much below 32 °F (0°C), or if it does, it's only for a short time. Air-source heat pumps can still operate quite efficiently at temperatures as low as 14 °F (-10 degrees Celsius).

Air weighs about 1 kg/m^3 and has low heat energy capacity. You need to process about 1m^3 of air per second to get enough heat for a small heat pump system. The air is abundant, and all we need is to use a fan to allow needed airflow.

Water is the best source of heat, as its heat capacity is about 3,500 times greater than that of air. However, this option is not often available to homeowners. To use it, you must have a lake, well or underground river near your house.

The Soil

The heat capacity of the soil itself is about half that of water. That's still an extraordinary amount, and we happen to have a lot of it. This huge store of heat beneath us can be used to our advantage with the help of heat pumps. The advantage is that the heat is still there even when it is extremely cold above. This is a challenge for any type of heating system. The soil tends to retain its heat, but when the demand for heat is exceptionally high and the "cold" penetrates from above, even the geothermal source can have problems at such temperature extremes. However, as we will see later, it is often acceptable to support the heat pump with alternative heating methods, such as gas boilers, since the total cost for these short extreme periods is only a small part of the total annual cost.

CHAPTER 2: OPERATING PRINCIPLE AND STRUCTURE OF HVAC SYSTEM

You don't have to know much about how HVAC systems work to own one, but a basic understanding is helpful before considering buying and installing one. If you decide to repair your HVAC system, you need to know and understand them.

HVAC systems are different. HVAC system collect thermal energy and transfer it from one place (for heat pumps it is usually a cold place, such as the ground, groundwater, or air) to another place (a room in a building) in the form of hot water or air. In case of air conditioning, we move take the heat from home air (this makes it colder) and transfer it outdoors of the building (outdoor unit).

Energy can neither be created nor disappear.

Energy can be transformed or transferred, but it can neither appear nor disappear. If something becomes colder, it is because it has "given" a certain amount of heat to something else, and this thing has become warmer by the amount lost by the first object.

For example, boiled water that got cold "gave off" heat to the room, and the room got warmer by that amount of heat.

Some modern gas boilers have an efficiency of about 95% (5% loss), which is very high. The efficiency is always less than 100%. For example, if we use 1 kWh of electric energy and convert it into heat energy with an electric heater, we get 950 to 990 W of heat energy. If we use a heat pump, which also uses 1 kWh of electric energy for its work and "pumps" heat from nature (ambient heat), we will get 3 to 4 kWh of heat energy on average. The efficiency of the heat pump can be between 300% and 500%.

This is what makes this technology so lucrative and interesting.

Part of the energy needed to operate the heat pump system (electricity) also enters the house as heat, but with small losses.

Most heat pump systems require electricity to operate. This makes this technology relatively environmentally friendly, as power plants normally use coal to generate electricity. Of course, over time, more and more electrical energy will be generated from renewable sources such as solar, wind, ocean, heat, etc. You can also install a photovoltaic system on your home to increase the amount of green electricity you use and increase your independence from energy prices and the power grid. In the future, most of the energy consumed by mankind will probably be generated in the form of electricity.

How does a heat pump transfer heat from heat sources?

The answer is simple: we need to create cold!

As you recall, heat moves from a high temperature to a low temperature until it becomes equal at two places.

How do you get heat from something that has a temperature of 23 °F (minus 5 degrees Celsius)?

To do this, we need to create cold that is below 23 °F, for example, 5 °F, and then bring them into contact with each other.

Now, thermal energy naturally begins to move from 23 °F to 5 °F degrees, and an object that is 5 °F begins to store heat that we can later extract using another law of nature (through gas compression) and use that heat to heat a room or entire building. Pretty simple, right?

We all have a heat pump to keep our food cold! It is a refrigerator. The refrigerator literally "pumps" the heat from the inside camera of the refrigerator to the back and releases the heat outside into the room.

Explanation: If the refrigerator cools its camera to, say, 32 °F, this means that 36 °F of heat is released to the back coil of the refrigerator, which can be heated up to 140 °F and slowly releases the heat into the room. The greater the temperature difference between the room and the back coil of the refrigerator, the more efficiently it works. The faster the heat is released to the room.

No heat was generated or destroyed - it was merely transferred from one place to another.

With heat pumps, for example, an air source heat pump, we don't take the heat from inside the refrigerator, but we collect the heat from a large amount of air outside the building and transport it inside the building, into the room, and there we give it to the room. In the case of cooling the house – the heat is gathered inside the building (from air) and transferred outdoors.

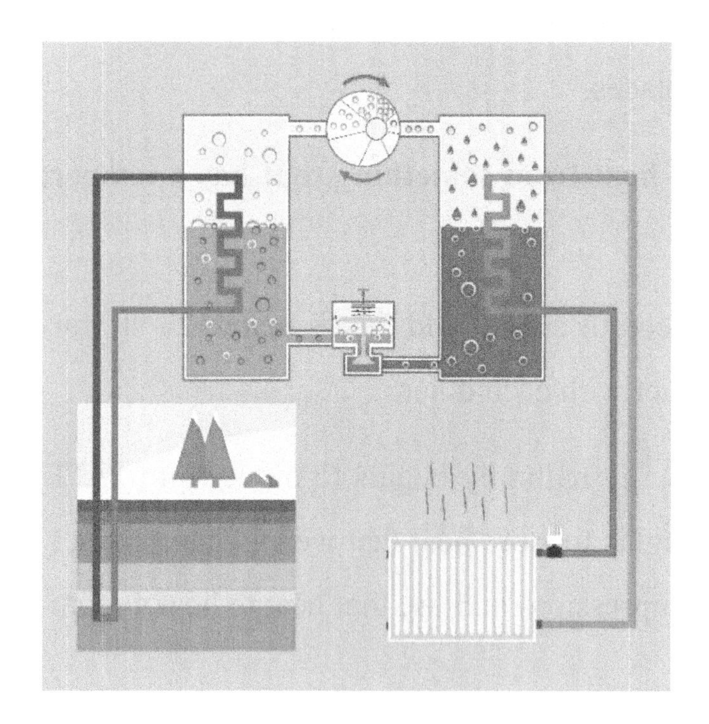

Now we understand:

32 °F or 273.15 K is the temperature at which water freezes (changing state), and below that temperature, there is a lot of heat (energy) that can be used.

Absolute 0 is equal to 0 Kelvin or -459,76 °F. This is the temperature at which there is no energy, and the molecules and atoms theoretically stop moving. Now let's look at the components of an HVAC system and what role they play in the system.

Structure of an HVAC system

An HVAC system consists of the following main components:

1. **Evaporator coil** – a component of the system designed to allow efficient absorption of heat from the environment (air). Contains inside a refrigerant with low pressure and low temperature (below ambient).
2. **Outdoor unit** – typical air conditioner outside unit which consists of several main components – compressor, condenser coil, and fan. Designed to release heat in case of cooling or absorb heat in case of heating the hose.
3. **Compressor** – used to compress gas at low pressure and temperature, thereby increasing the temperature and pressure of the gas to release heat to the environment.
4. **Condenser coil** – serves to release the heat that has been absorbed by the evaporator into the environment.
5. **The metering device (expansion valve)** is used to control the flow of refrigerant through the system and to control the heat transfer in the system. As the refrigerant passes through the expansion valve, the gas expands and reduces its pressure and temperature to the level necessary to absorb heat. The expansion valve is located in the refrigerant line between the condenser and evaporator, which are connected.
6. **Refrigerant lines** are used for the flow of the refrigerant in the system and connect indoor and outdoor units.

7. **The thermostat** is another widely used element of the system, which is used to control the temperature and operation of the HVAC system. Usually, the control panel is installed on the wall inside the building.
8. **Air filters** –they ensure air quality and are designed to remove dust and other particles from the indoor air that needs to be delivered to the building premises.
9. **Ductwork** – a network of ducts for transferring heated or cooled air to all areas of a building.

All these components are designed to form a closed thermal circuit (refrigerant loop). There is a cold and a warm or hot part of the circuit. Depending on the need, we can choose which part of the system we want to use. If we need to heat a building, we use the hot part, and if we need to cool the house, we can convert the hot part to the cold one. The operating direction of the system can be easily reversed, if necessary, because the condenser and the evaporator are very similar, if not identical.

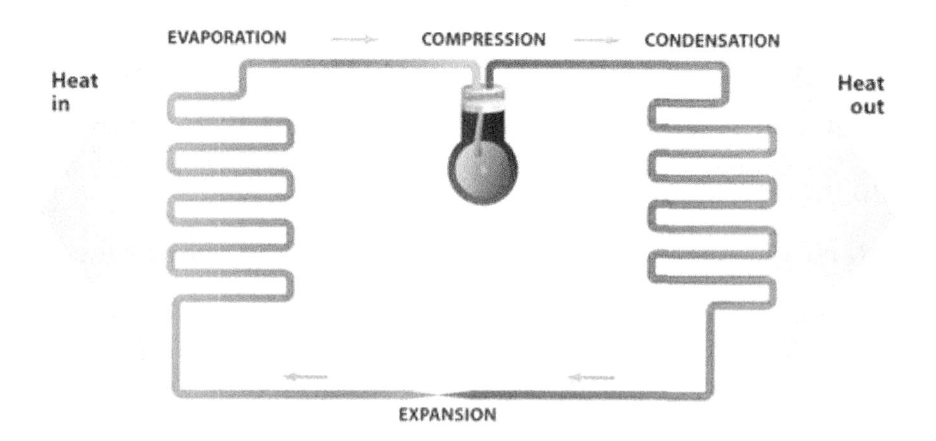

The compressor compresses the gas refrigerant and heats it. After this, it moves to the condenser coil where heat is released. To continue the cycle, the refrigerant gas that has condensed on the "hot" side of the cycle (condenser coil) must flow back into the evaporator coil. When the refrigerant goes through the expansion valve, the pressure drops, making the liquid colder again to absorb heat and evaporate into gas. So now we have a closed loop that can continue to run as long as we want.

The cold part of the cycle (heat absorption by evaporator coil)

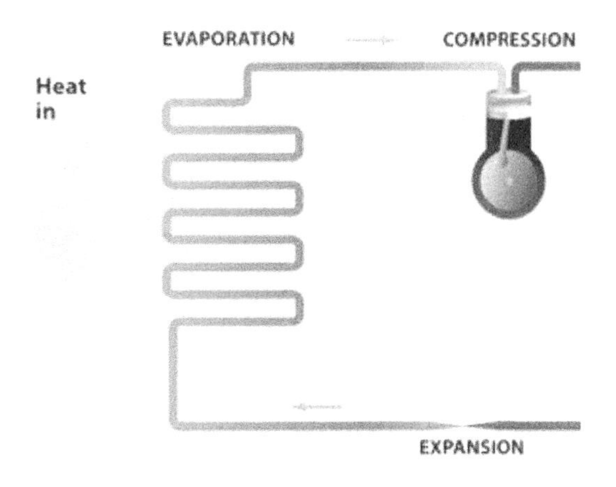

The hot part of the cycle (vapor compression and heat release)

When a gas is compressed, it heats up. Anyone who has tried to inflate a bicycle tire has noticed that the lower part of the pump, where the air is compressed, heats up. A simple heat pump can be built using the principle of compression and expansion of a gas.

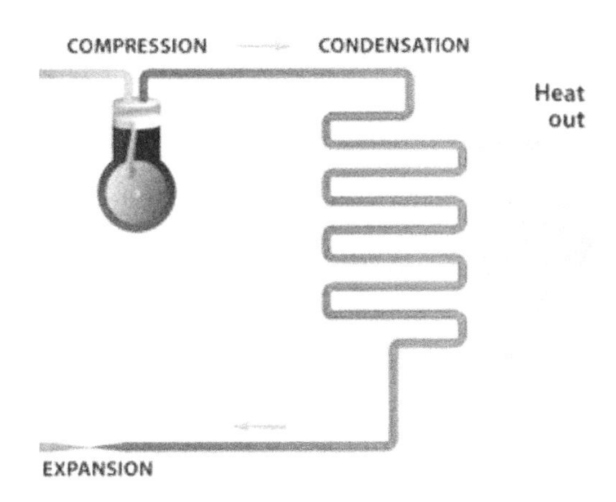

When it cools down to a level lower than the temperature of the heat source, it absorbs heat energy, and later the compressor of the heat pump or AC compresses the gas to a temperature higher than that in the building, whereupon the gas releases heat energy that warms the room. Again, in the case of AC, the heat is released outside in the outdoor unit by a condenser.

Evaporator coil

The evaporator consists of a series of tubes made of copper or aluminum, which are connected in series in the form of a coil. Aluminum and copper are materials with high thermal conductivity, which ensures efficient heat transfer. A coil provides a large surface area that increases the heat transfer rate.

It acts as a heat exchanger. Heat is transferred from the external environment to the refrigerant in the evaporator coil. The absorption of heat from the environment results in a change of state from liquid to gas. The evaporated gas has a slightly higher temperature than the liquid. *For efficient heat transfer, the evaporator is required constant circulation of water or antifreeze (at ground or water source systems) or air flow through its coil.*

Compressor

This element of the system is responsible for compressing the refrigerant vapor coming from the evaporator. As we already know, it increases the vapor pressure and temperature of the refrigerant. The compressor is powered by an electric motor. Some HVAC systems use variable-speed motors that have higher efficiency. To minimize friction of moving parts, most compressors have an oil pan in which lubricating oil is stored; other types require a lubricant to be added to the refrigerant. Gas/evaporator pressure can be increased up to 30 to 40 bar, depending on the type of compressor. The most common compressor used today is a scroll compressor, which is characterized by high efficiency, quiet operation, small size, and high reliability.

Compressors should be operated cyclically, i.e. not start up too often and not run for too long. In general, the compressor should not start more often than once every 15 minutes, ideally even longer, so that we could avoid wear on the equipment and to extend the compressor's service life. Continuous operation of the compressor can lead to overheating, which can cause damage.

Continuous work of the HVAC system (its compressor) may be a sign of a problem. For example, dirty filters or leaking refrigerant. Continuous work can also mean that the system is undersized, and the HVAC system cannot provide the required temperature in the building, and work constantly.

Most heat pumps have a single compressor, but there are other options:

Larger heat pump units can use a "tandem pair" of compressors. This is simply two compressors placed side by side and connected by common piping. Either one or both can run and provide 50 or 100 percent capacity.

Alternatively, a unit may have two separate cooling circuits. If one system fails, likely, the other will still function.

A more complicated application of a dual system is the "cascade", where one system follows the other. This is useful when high heating and cooling temperatures must be achieved. The disadvantage is that it is very complex and therefore this method is hardly suitable for residential buildings and is mostly used in commercial buildings.

Different compressor configurations can be used for HVAC systems - depending on the application, energy efficiency, and type.

There are several types of compressors used in HVAC systems:

- Single-stage and two-stage compressors.

- Reciprocating compressor (piston type).
- Rotary compressor.
- Centrifugal compressor.
- Scroll compressor (mostly used these days).

Condenser coil

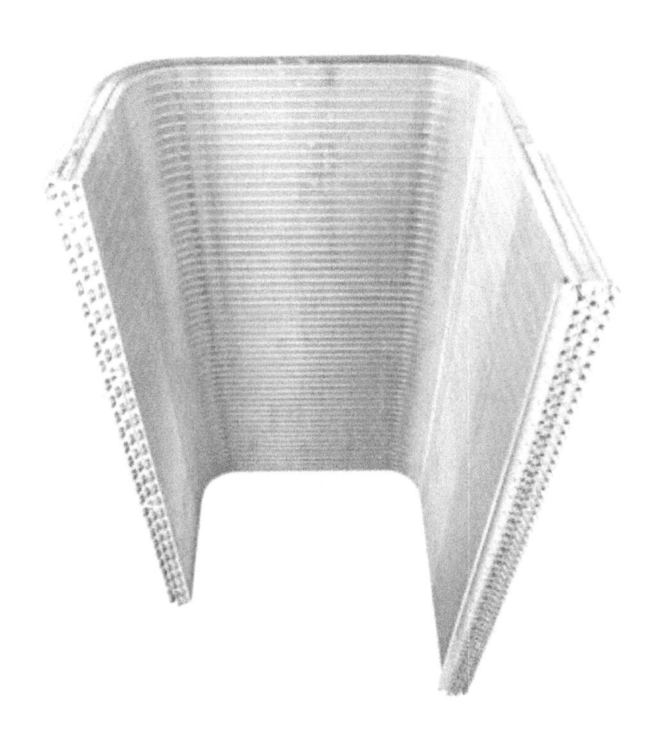

The condenser coil comes right after the compressor. The condenser works as a heat exchanger between the refrigerant, and airflow or circulating water. Here, the refrigerant vapor returns to its liquid state and gives off heat. The condenser consists of a copper or aluminum coil to which a series of aluminum fins arc attached. Just like the evaporator.

There are two main types of condensers used in HVAC systems: water-cooled and air-cooled. The first transfers heat to the circulating water, and the second to the air. The efficiency of water-cooled condensers is much higher than that of air-cooled condensers.

Expansion valve

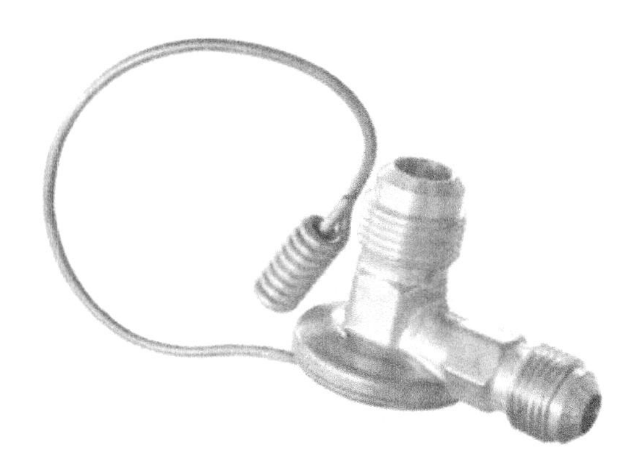

An expansion valve or thermostatic expansion valve is an expansion device through which the refrigerant flows before entering the evaporator. We have described its function before.

These 4 main components are basically the minimum required to create a heat pump (refrigerant) circuit.

Let's take one last look at the diagram and remember how it works in the closed loop.

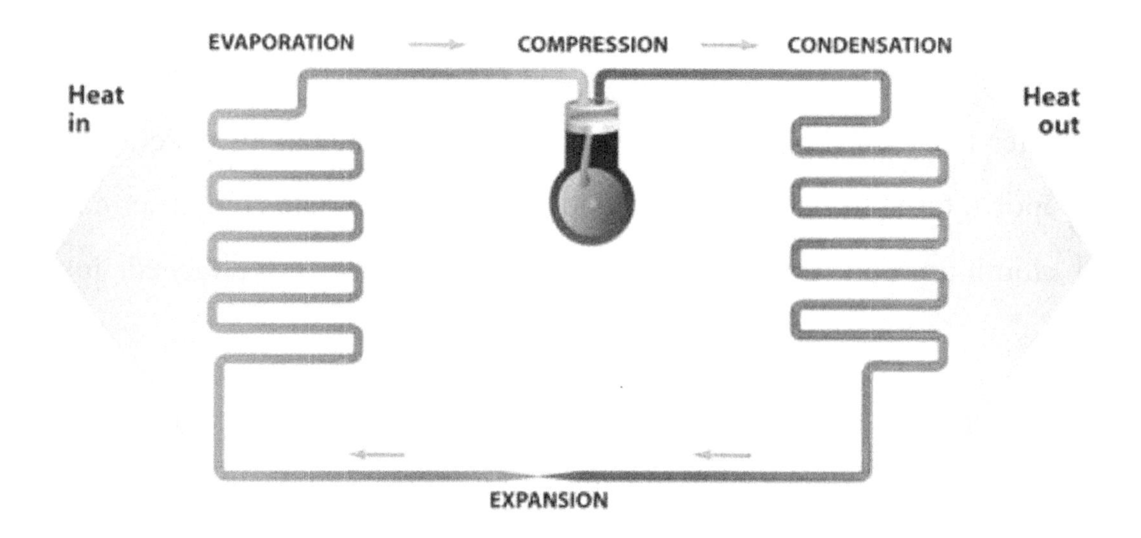

Reversing valve

It is used to reverse the refrigerant circuit and allows you to not only cool but also heat your home. The presence of this component distinguishes a heat pump from an air conditioner. Air conditioners do not have a reversing valve. In air conditioners cold is blowing inside the house and hot air is blowing outside the house. When heating – heat blows inside the house and cold blows outside (outdoor unit). **In reverse order.**

How do you find it? It's very simple - if you see one tube going into a system component and three coming out, it's a reversing valve. Often is located inside the outdoor unit.

Other elements that improve and ensure the proper operation of a heat pump

Fan

Fan for the indoor and outdoor unit - used to ensure sufficient airflow and, as a result, the desired heat transfer through the unit.

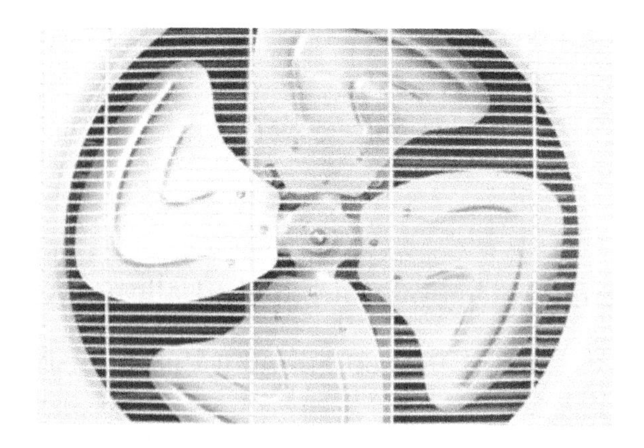

Circulation pumps - are used for the circulation of heated water in the building to ensure the heating of the entire building. We are already familiar with this component, as it is widely used in conventional heating systems. For example, if you have a gas boiler, a circulating pump is installed in it. With the air type of heating, a fan is used to create air circulation through the system (condenser coil) in case of house heating.

Filter-Dryer

It is used to remove moisture and other particles from the refrigerant that can enter the compressor and damage it or shorten its service life. The filter-drier is a cylindrical device with a filter medium and a desiccant that absorbs moisture from the refrigerant. It is installed in the refrigerant line between the evaporator (which absorbs heat and is cold) and the compressor.

The filter drier must be replaced together with the refrigerant when it is replaced. How often the refrigerant needs to be replaced depends on the operating conditions, the type of refrigerant used, and the size of the system. You can check this information on the website of the manufacturer or ask an HVAC specialist.

Pressure switch (high and/or low pressure)

This is a safety device designed to prevent refrigerant pressure from exceeding the normal range, and to keep the pressure within safe limits. The pressure switch turns a compressor on and off depending on the pressure value. If the pressure drops below the required pressure, the switch closes the circuit and turns on the compressor, which increases the pressure in the system. If the pressure is higher than required, the switch opens the circuit, and the compressor shuts off. This prevents damage to the compressor or system failures, such as refrigerant line damage and leaks.

Defrost control board and defrost thermostat.

This component is used in the heat pump and is usually located inside the outdoor unit of the heat pump. When the outdoor unit starts to freeze, the control board reverses the refrigerant flow (using a reversing valve) from heating the house to cooling and defrosts the outdoor unit. This usually happens when the temperature drops to around 0 degrees Celsius (32 °F) in winter. Sometimes the built-in electric heater is used to defrost the outdoor unit.

Disconnect switch (box)

A safety tool for disconnecting your outdoor unit from a 240 V power supply. Usually located on the side of your house near the outdoor unit. They can be circuit breaker type, plug, or lever type.

Electrical whip

The pipe contains all the electrical wires that run from the disconnect box to the outdoor unit.

Refrigerant Line Set

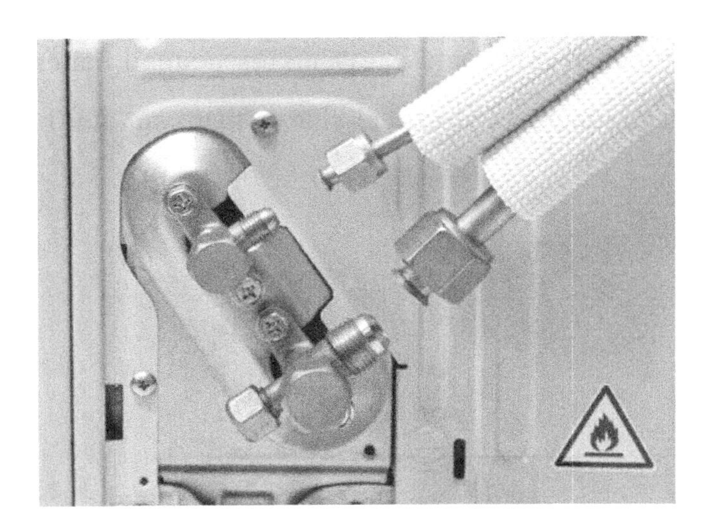

Two pipes (a thick and a thin one) leave the house and enter the outdoor unit. The thicker one is called **the suction line** and the thinner one is called **the discharge line**. The thicker pipe (suction line) always has insulation on it.

Service valves

Connectors (access ports) for connecting pressure gauges and checking the pressure level in the refrigerant system. We need to first connect the pressure gauges, open the valve cap, and then open the valve to be able to measure the pressure. Maintenance of refrigerants and their lines is not recommended for beginners, it is something that should be done by a trained technician when it involves refrigerant.

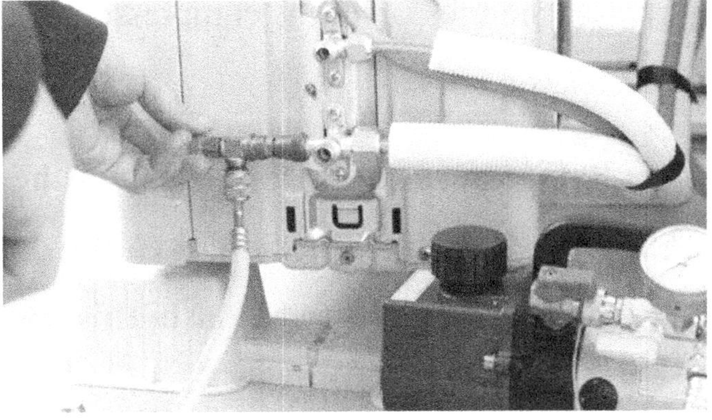

Capacitor

It is usually cylindrical in shape and is essentially a small battery used to start and run an electric motor. They accumulate an electrical charge that can later be given to the system (fan and compressor motor).

Run capacitor. It is used to create an artificial second phase for single-phase motors, which helps to spin up the motor (the rotor starts to rotate). Once the rotor starts to spin, the interaction between the rotor and stator keeps the rotor spinning.

Hard start capacitor. Helps to create a very high amperage for motors which makes the process of starting a motor very fast and easy. Without it, the motor would run for a long time with high levels of amperage which is bad for electrical components and wires. Prolonged high current causes heating. A hard start capacitor increases the lifespan of the compressor and is used only to help the compressor start-up (first second or so).

Before performing any work on the wires inside the unit, we must discharge the capacitor (the capacitor is charged and may cause an electric shock). How to do this will be discussed later in this book.

Contactor

It is a control relay that controls when to send power to the outdoor unit and start the HVAC system to cool or heat your house. When the temperature in the house rises above the temperature set in the thermostat, it supplies power to the contactor, which starts the outdoor unit to start the cooling process. And when the house cools down to the desired temperature, the thermostat stops working, supplying power to the contactor, and the unit turns off.

To get 240 V to operate the outdoor unit, we need to combine two wires, each with 120 V. These two wires are connected to the contactor.

The contactor consists of an electric coil that is powered by 24 V (27 V can be measured) coming from the home thermostat or furnace control board. This electric coil generates a magnetic field that moves a plunger in the contactor, which closes the 240 V electrical circuit that powers the outdoor unit (compressor and fan). In the image above, the two wires coming from the side (right) provide 24 V power.

Refrigerant

The working fluid in an HVAC system is called a refrigerant and is used in every refrigerator, heat pump, and air conditioner.

Requirements for the refrigerant:

- non-corrosive
- non-toxic
- non-flammable
- can be mixed with oils for compressor lubrication.

In the early days, the refrigerant industry experienced rapid growth, and **(CFC) Chlorofluorocarbons gases (R-12)** were considered ideal and widely used.

They were inert, non-toxic, and had ideal thermal properties. Unfortunately, this liquid also had a disadvantage that was not recognized and acknowledged at the beginning. It was discovered that these gases destroy the ozone layer of the atmosphere, which is needed to filter the sun's harmful ultraviolet rays.

Hydrochlorofluorocarbons (HCFCs) - are less damaging to the ozone layer than CFCs, but are still harmful and therefore not widely used. An example of such a refrigerant is **R-22**.

*One of the most common refrigerants used today is **hydrofluorocarbons (HFCs)***, which do not affect the Earth's ozone layer, but are still considered a gas with an impact on global warming (greenhouse effect). There is still some debate about the use of this refrigerant. Examples of refrigerants - **R-410A, R-32, and R-134a.**

Today, there are strict regulations for the use of refrigerants and the refrigerant gases must not be released into the atmosphere. When disposing of old equipment, in most cases the law requires that refrigerant be collected and recycled (but in your area it can be different).

Another newer type of refrigerant that is widely used and has a low impact on the ozone layer and global warming is **Hydrofluoroolefin (HFOs)** such as **R-1234yf and R-1234ze**.

Hydrocarbons are relatively environmentally friendly refrigerants (natural refrigerants). They are very good, but have one drawback: they are flammable, so you should be very careful when dealing with them.

Many household refrigerators use these hydrocarbons: R600A, which is isobutane, propane (R-290), ammonia (R-717), and carbon dioxide (R-744). HVAC units usually have a relatively large fill volume, usually 1 liter. This is considered somewhat hazardous and is therefore usually only used for equipment that is located outside the home.

In most cases, refrigerant handling should be performed by HVAC professionals to ensure safe refrigerant disposal.

Condensate drain line

It is a pipe that is used for running the condensate outside of the house. When your heat pump or air conditioner is in cooling mode, the heat exchanger inside your home becomes an evaporator (cold to absorb heat from the air) moisture from your home's air begins to condense on the evaporator and flows down into the drainage line.

Condensate overflow switch

It is essentially a sensor that reacts to water leaks. If the condensate drain line becomes clogged, it shuts off the power to the HVAC system to prevent leakage of condensate inside the house.

Inverter motors

Inverters are used to control the speed of the compressor motor of an HVAC system (outdoor unit). This makes the room temperature more stable than in the case of a system without an inverter when it is only turns on and off, which leads to temperature fluctuations in the building of about 5-7 degrees, which is not very pleasant for residents.

HVAC systems with inverter compressors are more energy efficient, run quieter due to lower motor speed, can heat, or cool a room faster than a system without an inverter, and reduce compressor wear.

Energy efficiency - coefficient of performance (COP)

COP equals energy in the form of heat that we get from a heat pump (Php) divided by the energy (electricity in kWh) we need to operate a heat pump (Pb).

COP = Pwp / Pb

For example:

The HVAC system consumes 1 kWh of electrical energy, and we get 3 kWh of thermal energy to heat a building.

COP = useful heat output / electrical power consumption = coefficient of performance (COP)

The equation for calculating the efficiency:

COP = 3 kWh / 1 kWh = 3

The COP in this case is equal to 3, this means that we get three times more energy than we spend.

The energy efficiency of a heat pump strongly depends on the application and the temperature difference (dt) between the heat source (ambient temperature) and the flow temperature (heat transfer medium – water or air flowing through coils).

For example, if heat is extracted from a river or a well (water) at a temperature of 6°C and the heat pump begins to heat the heat transfer fluid (air or water that heating the house) to 53°C (dt = 53 - 6 =47°C), the coefficient of performance (COP) is probably about 2.7. However, if the heat pump were to heat the heat transfer fluid to 35°C instead of 53°C (dt=35 - 6 = 29°C), the COP would be 3.8 or more. This is because the system has to do less work to heat the heat transfer medium to 35°C than it would if it had to heat it to 53°C.

In other words, **the larger dt is,** the less efficient the HVAC system (in this case heat pump) works (the COP number becomes lower), and **the lower the dt is, the more efficient the heat pump works** (COP becomes higher).

The COP is an important basic coefficient that provides information about how well an HVAC system is working at a given time.

Temperature differences between the temperature of the heat source and the temperature of the refrigerant that absorbs the heat from the environment. In other words, the warmer the source and the cooler the sink (refrigerant), the better.

For cooling (AC applications) – the lower temperature outside the building the better. Because the condenser gives off heat fast and efficiently. *In this case, the larger the dt, the better.* If it is hot outside, the condenser heat exchanger will not transfer heat well – and the system works less efficiently. *COP number becomes smaller.*

Another factor that affects efficiency is the ***amount and type of refrigerant in the system***. If there is too much or too little, the system will not operate efficiently. Also, some refrigerants are more efficient at transferring heat than others.

The type of compressor used in the system is also an important factor affecting the COP value of the system. Some compressors use less electrical energy to compress gas than others, resulting in higher efficiency.

The low circulation speed of water or air can cause poor heat exchange in the system (between the condenser and the air or water circulating through the system), reducing the COP value.

For example, if the system is ***not properly maintained,*** dirty filters are not changed, the heat exchanger (condenser and evaporator coils) is not cleaned, and other factors can reduce the efficiency of the system.

Standard Methods for Displaying the COP

You can easily find the COP number on the product label, where the COP is given as a ratio, e.g. "COP 3" or "COP 4.4". The HVAC system manufacturers provide product documentation showing the dependence of the COP at different operating temperatures.

If you are interested in any HVAC system model and want detailed information, you can view the documentation on the manufacturer's website.

There are 2 types of COP that can be displayed on the HVAC system:

1. For heating (shows the efficiency of heating)
2. For cooling (shows the efficiency of cooling)

For a typical ground source heat pump:

COP = 3.8 (B0/W40)

The indication "B0" refers to the temperature and the type of liquid (B – Brine) when it comes back from the ground. The indication "W40" refers to the flow temperature of the heated water.

For a typical air source heat pump:

COP = 2.8 (A2/W45).

The "A" means air and that the air enters at 2°C (its external temperature), and W45 means that the heated water (heat carrier) enters at 45°C (flow temperature).

For US-manufactured HVAC systems, this might be a bit different, and the temperature will be shown in Fahrenheit.

The COP is a useful indicator of efficiency at certain operating temperatures. However, for air-source heat pumps, heat source temperatures vary throughout the year, so high COP values can be expected in fall and spring, and low values in winter.

The COP values for heating in air-source heat pumps can be changed like this:

Outdoor temperatures	COP
55 °F	3.5
41 °F	2.7
23 °F	2.1
14 °F	1.5

The COP values are different for the different heat pump models. The above figures serve as an example.

Annual performance factor (APF)

The APF is a measure of the total thermal energy delivered by the HVAC system during the year divided by the total electrical energy delivered during the year.

To show how efficiently the heat pump will work during the year, the **annual performance factor (APF) is used.**

For example, if the temperature of the heat source (air) varies during the year between -10 °C and +15 °C, which corresponds to the temperature range in which we use a heat pump for heating. The APF indicates the efficiency of the system more accurately than the COP, which indicates the efficiency for specific conditions and temperatures.

APF = Total annual thermal energy provided / Total annual electrical energy consumption (for heating)

For example:

Total annual consumption of electrical energy = 5000 kWh

Total annual thermal energy provided = 16000 kWh

APF = 16000 kWh / 5000 kWh = 3.2

The higher the APF, the more efficient and economical the heat pump is. It is highly dependent on the year-round ambient temperature (weather) and the thermal insulation of the building. The price of electricity in your region is also a very important factor influencing the economic use of heat pumps.

Advantages and disadvantages of the various heat sources

Source	Disadvantages	Advantages
Air	Heat pumps have low efficiency and can be troublesome at low temperatures. Low amount of energy in the air, a lot of air is needed to collect and transfer enough heat for use, especially when it is cold.	Easy to install. Abundant and easy to use without additional preparation.
Water	The source must be located near the building.	The efficiency is enormously high than in air systems. For this reason, the COP number is also very high. Can be cheap to use. Stable energy output, even at very low outdoor temperatures.
Soil	Expensive and not often required soil space is available. Difficult to install.	They can be very energy efficient. High COP and APF numbers. Stable heat output, even at very low outdoor temperatures.

CHAPTER 3: HEATING AND COOLING THE HOME HVAC SYSTEM CONFIGURATIONS

Heat Pump System Types

Every heat pump has two sides: the source and the sink.

When describing heat pumps, the source is often mentioned first, followed by the sink, with a "to" in between. Water-to-water heat pumps, for example, describe a system that extracts heat from water (well, pond, or lake) to produce hot water for radiators or radiant floor heating.

Thus, the main types of heat pumps are:

- Air to air.
- Air to water.
- Water or brine to water.

Air-to-Air Heat Pumps (or simple AC)

Let's talk about air as the easiest to use and most common heat source. It is present everywhere and available everywhere. The ability of air to store heat is low. To extract a significant amount of heat from it, we need to pass large volumes of air through a heat exchanger (evaporator coil) to "extract" sufficient heat energy.

The air-to-air heat pump consists of 2 main blocks, an outdoor unit, and an indoor unit (evaporator). They can be mounted on the wall or on the ground next to the building.

Outdoor unit that can be mounted on the wall.

Outdoor unit for floor installation

Air-sourced heat pumps can be found in every city and in all hot countries. They are commonly known as air conditioners for cooling buildings. But air heat pumps have reversing valve built in, which allows the system to be switched any time there is a need for heating in the building. I believe that all air conditioning units should have this option if you live in an area that gets cold from time to time. Technically, the construction of a heat pump is the same with only a few additional components.

The air flows through the outdoor heat exchanger (evaporator coil) and flows to the atmosphere a few degrees colder. In this mode, heat is extracted from the outside air and this energy is transferred to the building for heating.

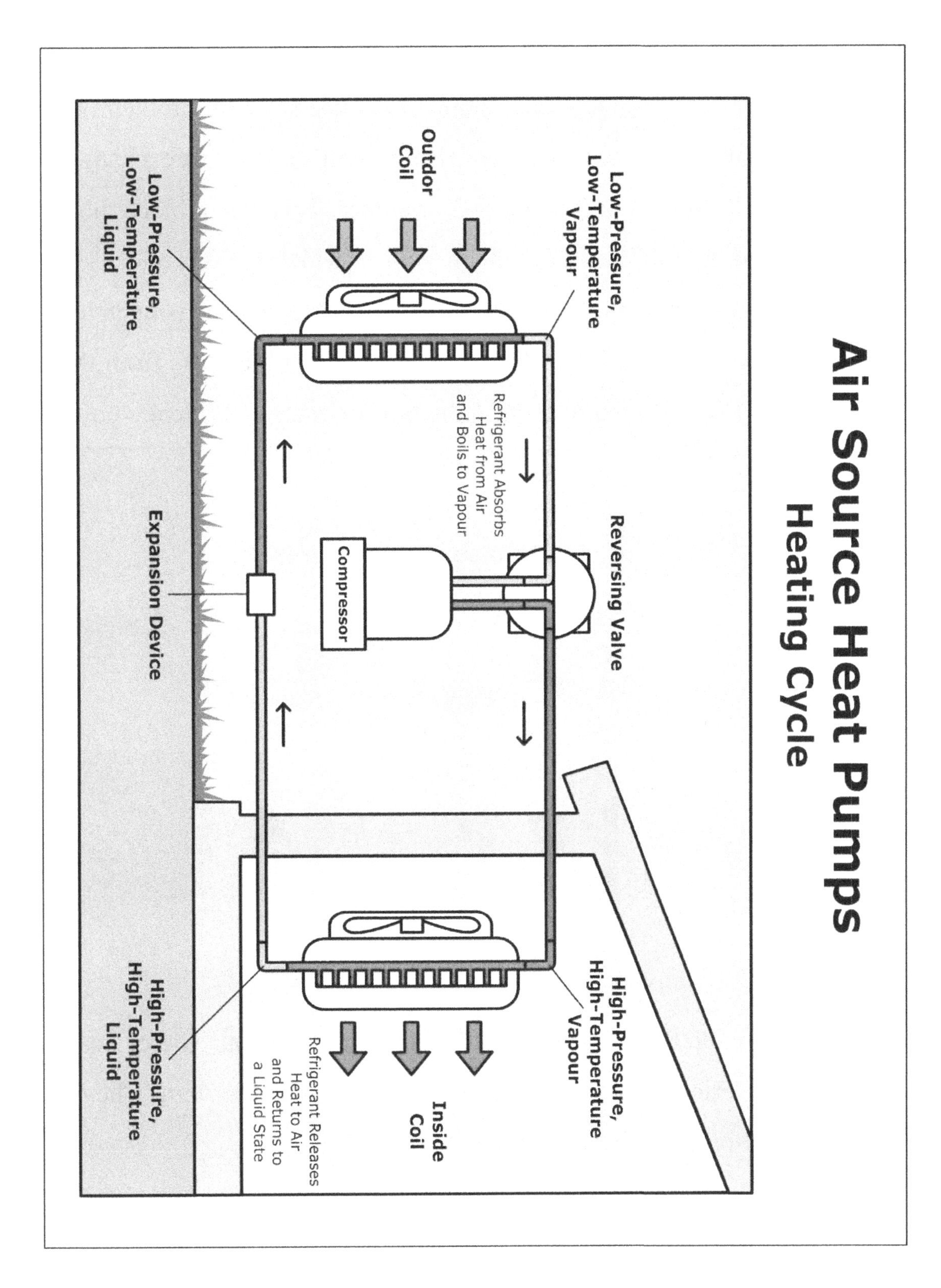

However, there are also many air heating systems that are designed specifically for heating. These are usually air-water systems that are **much more energy efficient.** We will talk about this in more detail later.

Outside air continuously flows through the outdoor unit, and when the colder air leaves the heat exchanger, the water begins to condense. One problem with any air-source heat pump, however, is that condensed water on the heat exchanger fins turns to ice when the incoming air is about 6°C or colder. This would block the airflow; the problem is fixed by a defrost mechanism built in which melts the ice. This can happen once an hour or two, but the energy required to melt the ice is relatively low. The defrosting process can take up to five minutes, and the total loss from the necessary defrosting could result in an average reduction in energy efficiency on the order of only 5 to 10 percent.

The outdoor unit is frozen

Because it is exposed to the elements outside the building, the air-source heat pump system is likely to deteriorate more quickly than a geothermal system. Cheaper systems tend to be less well-protected from heavy rain.

This type of system is almost not used nowadays, as it is not very efficient compared to other heat pump systems. The coefficient of performance (COP) of such systems is between 2 and 3, but realistically rarely exceeds 2.5. Another significant disadvantage of air-to-air heat pump systems is that they are only suitable for heating, and for the preparation of hot water, another device is needed, such as an air-to-water heat pump or a conventional electric boiler. Such systems are useful where hot water is already available and only heating is needed, even if the apartment is very small (1-room apartment) and has a low heat demand. The price of such small systems starts from 1000 euros, which makes this system affordable for almost everyone, and the installation process is very simple, resulting in low installation costs. In some cases, you can even do the installation yourself, because some manufacturers offer DIY-friendly air source heat pumps.

Air-to-Water Heat Pump

Air-to-water heat pumps can be used both for domestic hot water and for heating the apartment, which makes this technology a better solution for multi-room apartments. We can connect this system to standard radiators that are commonly used or to a floor heating system, which makes heat transfer more efficient due to the larger heat transfer area and lower temperature required.

The structure and principle of operation of this system are almost the same as air-to-air heat pumps. The only difference is that the heat in the building is transferred from the refrigerant to the water which circulates through the water circulation pump in the system (radiators and floor heating). For efficient operation of the system, it is necessary to keep the operating temperature of the water within 104 °F (35-40 degrees Celsius), and then the coefficient of performance (COP) will be between 3 and 5. If we bring the operating temperature of the water to 158 °F (70 degrees Celsius), which is necessary if we decide to use radiators, then the APF will be between 2 and 2.5, and it is much lower than if we use an operating temperature of 104 °F (35-40 degrees). Knowing this, it is clear that we should avoid high operating temperatures for heat pumps.

Underfloor heating and radiators

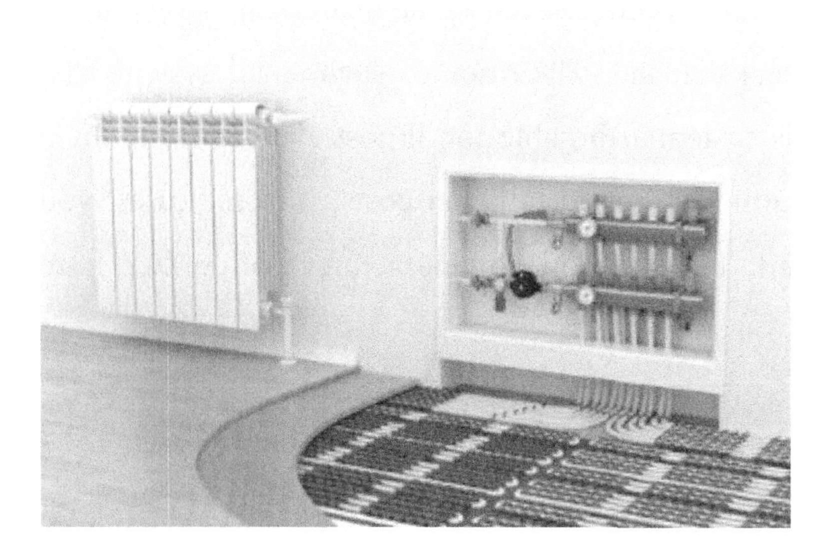

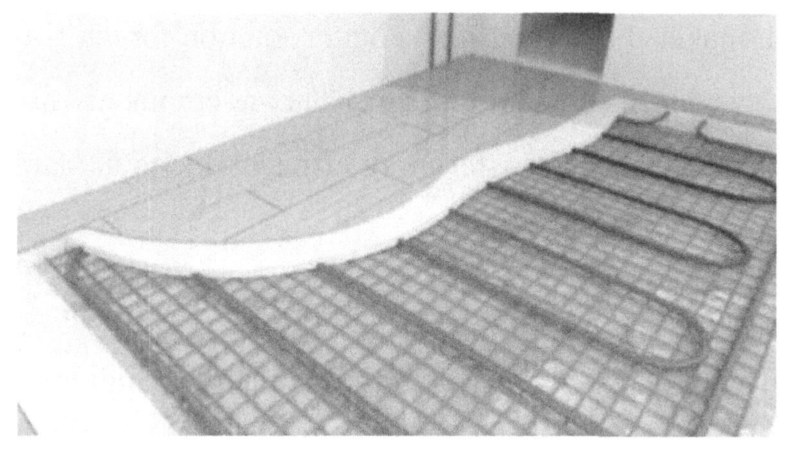

Another disadvantage is that the outdoor unit can freeze at low outdoor temperatures, which reduces efficiency and interrupts work. The efficiency of such a system is about 300% (COP =3). Under ideal conditions, some models can achieve a COP of up to 5. The annual performance factor of a water-to-water heat pump system is between 3 and 4. The price of such systems starts at about 10,000 dollars, which is not much compared to geothermal heat pumps. If you take into account the installation, it will cost the homeowner between 15000 and 20000 dollars. Financing is usually not required for this type of system, and if it is, it is not very high, which makes this type of heat pump a very attractive and affordable option. Compared to simple air-to-air heat pumps, the installation costs for this system are much higher and it is more complex, so it is not suitable for do-it-yourselfers. The efficiency of this system is also very dependent on outdoor temperatures and requires supplemental heating support at very low temperatures. Noise from the outdoor unit is also a factor to consider, and in some cases, additional soundproofing may be required.

To be honest, this type of heat pump is probably best for most people. We can effectively use this system for most of the heating season, and only need additional electric or gas heating when temperatures are extremely low. When the outside temperature drops below 10 degrees Celsius, we probably need additional heating (5 to 10% of the time during the heating season). To use this type of heat pumps, we do not need to do anything special outside the building, such as digging trenches. This strategy of home heating is very popular and has proven itself in Scandinavian countries. This strategy is used in Sweden, where it has proved to be practical and economical to install a heat pump that provides only two-thirds of the expected maximum heat demand of a building.

The outdoor unit of the heat pump should not be installed too close to the wall, as this reduces the airflow and thus the efficiency of the system.

When installing an air-to-water heat pump, maintain a distance of at least 6.5 ft (2 meters) from surrounding properties. Depending on the power of the heat pump, make sure that the power supply line has a sufficiently large cross-section to ensure the safe operation of the HVAC systems. If there are neighbors nearby, it is recommended to increase the distance so as not to disturb the neighbors. Modern heat pumps produce ventilation noise in the range of about 60 decibels (dB), which is very quiet and hardly noticeable. Sometimes it is useful to measure the noise level.

Carefully seal the penetration point through the outer wall of the house (through which the outdoor and indoor units of the heat pump are connected by cables and refrigerant pipes). All pipe connections to the outdoor and indoor units must also be insulated. A pipe for condensate drainage should be connected to the outdoor unit.

Geothermal HVAC Systems (Brine-Water Heat Pump (BWHP))

Now we can consider one of the most expensive, but at the same time one of the most efficient types of heat pumps - brine-water heat pump. The cost of such a system starts at $20,000 without earthworks. The reason for its high efficiency is the relatively stable and high temperature of the heat source - the soil. The installation of this type of heat pump requires a large plot of land near the house, where earthworks can be carried out, which for many of us is not possible (simply because there is no or not enough land) and brings additional difficulties, such as obtaining permits and the risk of damage to underground infrastructure (water or gas pipes or power lines).

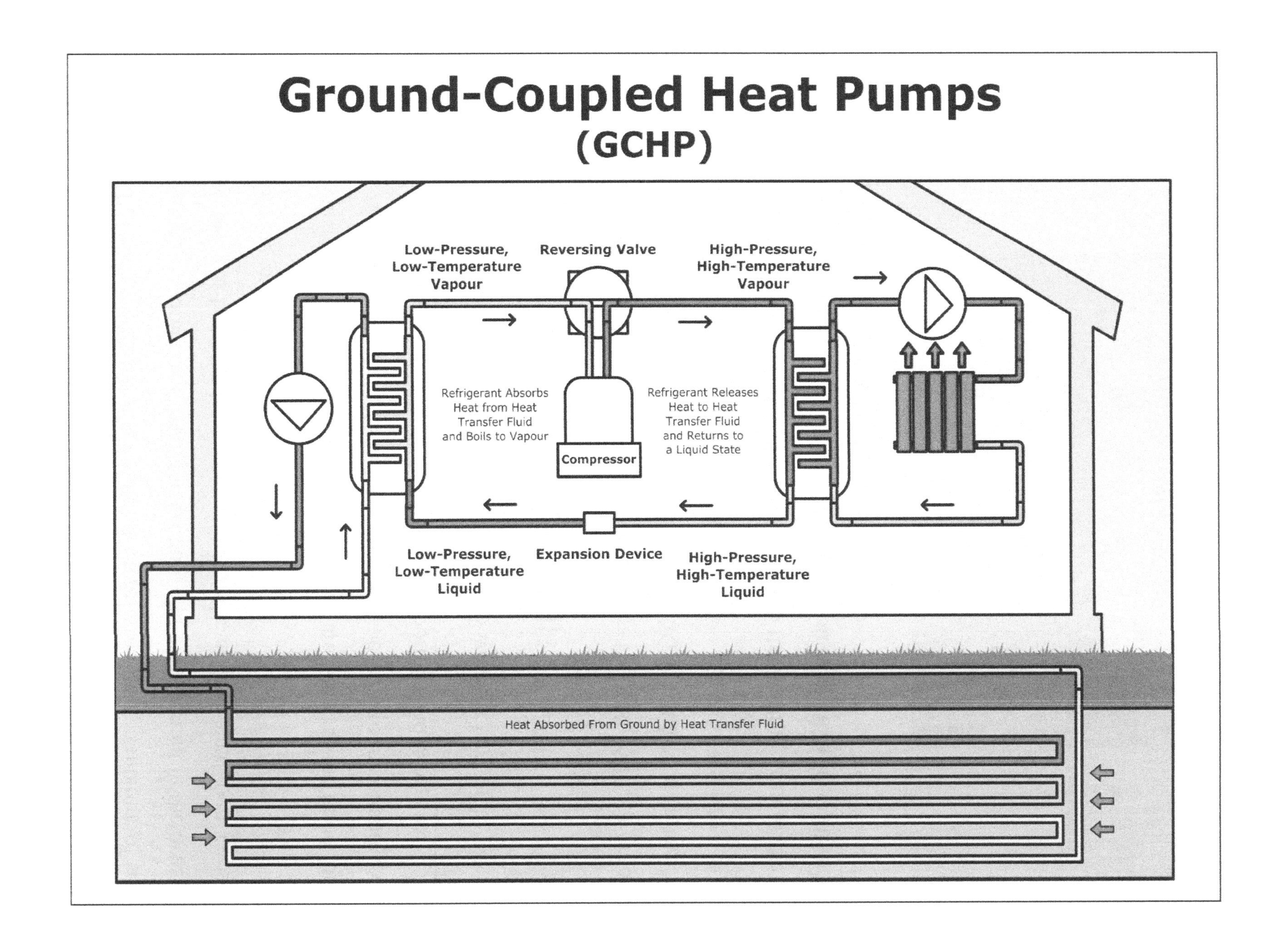
Ground-Coupled Heat Pumps
(GCHP)
Low-Pressure,
Low-Temperature
Vapour
Reversing Valve
High-Pressure,
High-Temperature
Vapour
Refrigerant Absorbs
Heat from Heat
Transfer Fluid
and Boils to Vapour
Compressor
Refrigerant Releases
Heat to Heat
Transfer Fluid
and Returns to
a Liquid State
Low-Pressure,
Low-Temperature
Liquid
Expansion Device
High-Pressure,
High-Temperature
Liquid
Heat Absorbed From Ground by Heat Transfer Fluid

At a depth of 4.9 - 6.5 ft (1.5 to 2 m), there is a surface collector (see the image), which is a long flexible pipe closed in a circuit. A circulation fluid, a mixture of salt and water (brine), is used to extract heat from the ground, and a refrigerant in the heat pump (evaporator) absorbs heat from the brine circulating in the underground loop, constantly drawing heat from the ground.

Sometimes ethanol or other liquids that do not freeze are added to the brine for safety reasons. Freezing of liquids can cause work stoppages and, more importantly, cracks and liquid leaks into the environment. Repairs are then laborious and expensive.

The APF number of such a system is in the range of 4 to 6 which is very high.

Heat Yield of the Soil

It indicates how much heat energy we can extract from 1 m^2 (10.76 ft^2) of ground area. Normally, for every 1 to 2 m^2 (from 10 to about 21.5 ft^2) of outdoor area, we can extract enough heat energy to heat 1 m^2 of living area.

Ideally, the land area should be 2 times larger than the heating area of the home (to ensure that the soil does not freeze). If your living area is 1076 ft^2 (100 m^2), you will need around 2152.7 ft^2 (200 m^2) of land area.

We must make sure that no trees grow near the flat collector, because tree roots can damage the pipes. There is also a risk that the tree will dry out if the soil and roots start to freeze. After you have buried the flat collector, do not plant bushes or trees on the surface. It is better to plant a regular lawn.

If the homeowner does not have enough space for an area collector, he resorts to drilled wells or so-called **"earth baskets" or trench collector systems.**

The depth of the well can be up to **656 ft (**about 200 meters), but in most cases, the depth is about **328 ft** (about 100 meters). The temperature in such wells can reach from 10 to 15 degrees Celsius (50 to 59 °F). For each meter drilled we can get about 0.05 kW of heat.

On average, the temperature increases by 1 °C (1.8 °F) every 33 meters. In other words, every 100 meters (328 ft) the temperature rises by 3 °C (or 5.4 °F).

For example, at a depth of 500 meters temperature will increase on:

(500 m) 5 x 3 °C = 15 °C or **about 27 °F more than temperature in the soil above**.

Since 1 degree Celsius is equal 1.8 °F (1.8 x 15 = 27 °F)

The temperature at this depth is about 20 °C. As already mentioned, the temperature at a depth of 2 meters varies by 5 - 15 °C depending on the season. This heat is absorbed and stored by the earth through solar radiation.

The ratio between the cost of drilling (more than 200 meters) and the increase in efficiency of the heat pump makes it economically unviable due to the high cost of drilling. Instead of a single deep well, 2 or 3 100-meter wells are drilled in practice. A specialist must determine whether the required amount of heat is possible for a particular soil. The thermal conductivity of the soil can vary from place to place and depends on the type of soil. Therefore, a consultation with a brine-water heat pump specialist is advisable and appropriate.

If we want to extract 10 kW of heat, the calculation of the drilling depth looks like this:

Depth of one well = 10 kW / 0.05 kW per meter = 200 meters (656 ft)

However, for safety reasons, a safety factor of 1.3 (30%) is used:

200 x 1.3 = 260 meters (853 ft)

We can also drill 2 wells at 130 meters (260/2) or 3 wells at about 90 meters (260/3).

Let's move on to another type of heat pump that uses water as a heat source - the water-to-water heat pump.

Water-to-Water Heat Pumps (WWHP)

This type of heat pump is often referred to as a groundwater-to-water heat pump because it uses groundwater as a heat source. It is the least common type of heat pump because the heat source is rarely available.

There are two most common ways to connect to a heat source (water). The first is to drill two wells (extraction well and absorption well) to the depth of the underground water layer, and the second is to use a river or pond near the house. The principle of operation is the same as that of a ground source heat pump. The cooled water that we have taken from the extraction well (from which the heat pump has already extracted heat) is pumped to the absorption well, where the water is heated again by ground. When using groundwater, the chemical composition of the water must be considered. Chemically contaminated water can lead to rapid corrosion or the formation of deposits on the pipe walls, which in turn can cause pipe blockages. An annual performance factor (APF) of up to 6 can be achieved. This is because water has a very high heat capacity (ability to absorb large amount of heat) and thermal conductivity (high speed of heat transfer) compared to other heat sources. This type of heat pump is suitable for new and old buildings (badly insulated) due to its high efficiency.

To generate 1 kW of heat energy from water, we need a water flow rate of about 52.8 gallons (200 liters) per hour. To generate 5 kW of heat, we need a water flow rate of 264 gallons (1000 liters) per hour. The depth of the boreholes can be up to 328 ft (100 m), depending on the situation. At the beginning you need to conduct test drilling, the cost of test drilling starts from a few thousand dollars. The price of a small drilled well with a length of about 20 meters (65.6 ft) can be about $4000 or more. You also need a permit from the municipality (water authority) to drill a well.

There is a risk that the groundwater level will drop at some point in future, which could reduce the efficiency of the heat pump. It is also possible that the well will run dry for an unpredictable reason. You should take this risk into account.

The following pictures show some other ways to use water as a heat source. Using circulating water in a pipe loop and water-to-water heat absorption in a well or pond. These methods have several advantages, such as using the same water and the fact that the evaporator coil is not in contact with groundwater. No risk of damage to the evaporator coil or any other part of the system (corrosion or deposits on the pipes).

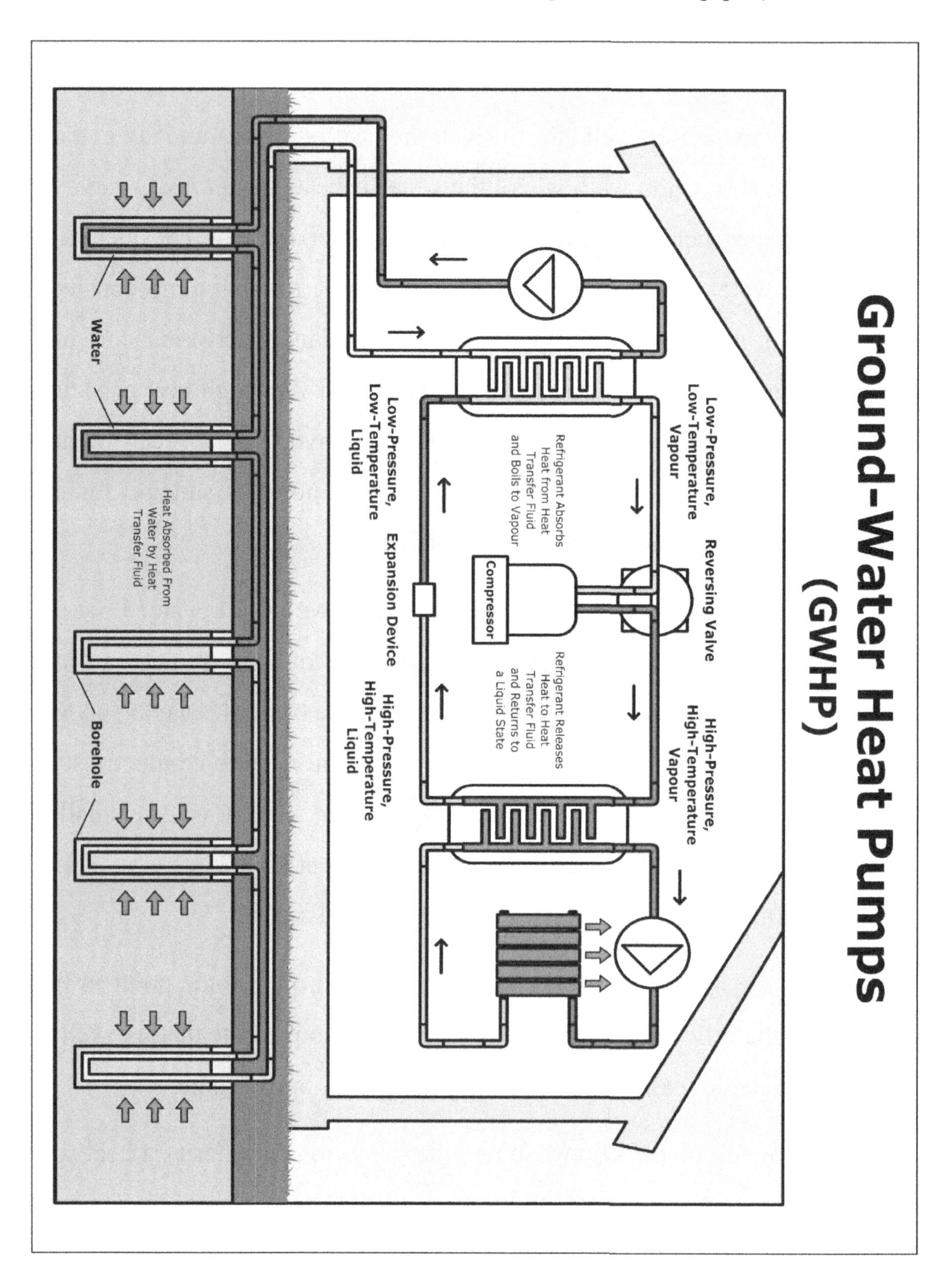

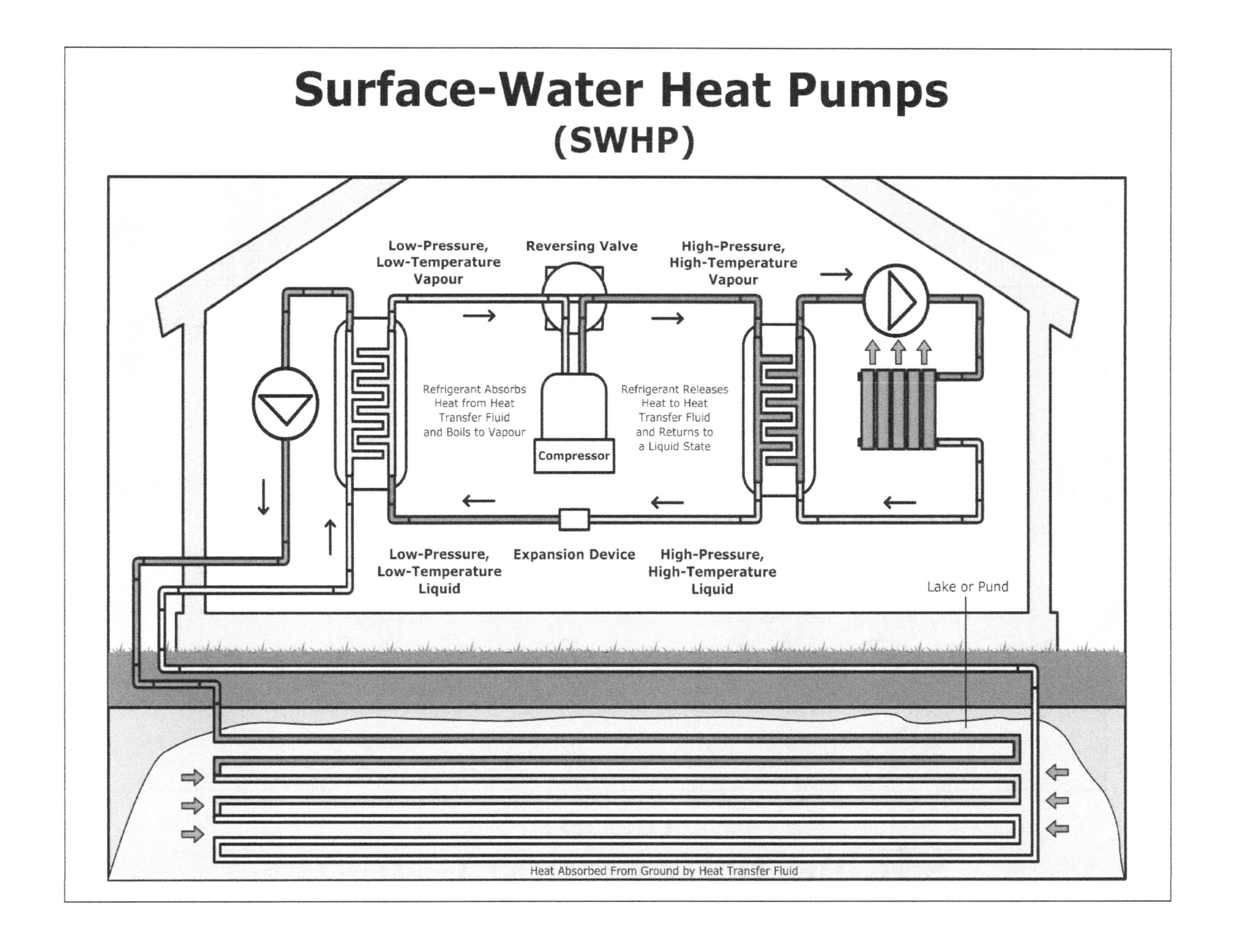

Surface-Water Heat Pumps
(SWHP)
Low-Pressure,
Low-Temperature
Vapour
Reversing Valve
High-Pressure,
High-Temperature
Vapour
Refrigerant Absorbs
Heat from Heat
Transfer Fluid
and Boils to Vapour
Compressor
Refrigerant Releases
Heat to Heat
Transfer Fluid
and Returns to
a Liquid State
Low-Pressure,
Low-Temperature
Liquid
Expansion Device
High-Pressure,
High-Temperature
Liquid
Lake or Pund
Heat Absorbed From Ground by Heat Transfer Fluid

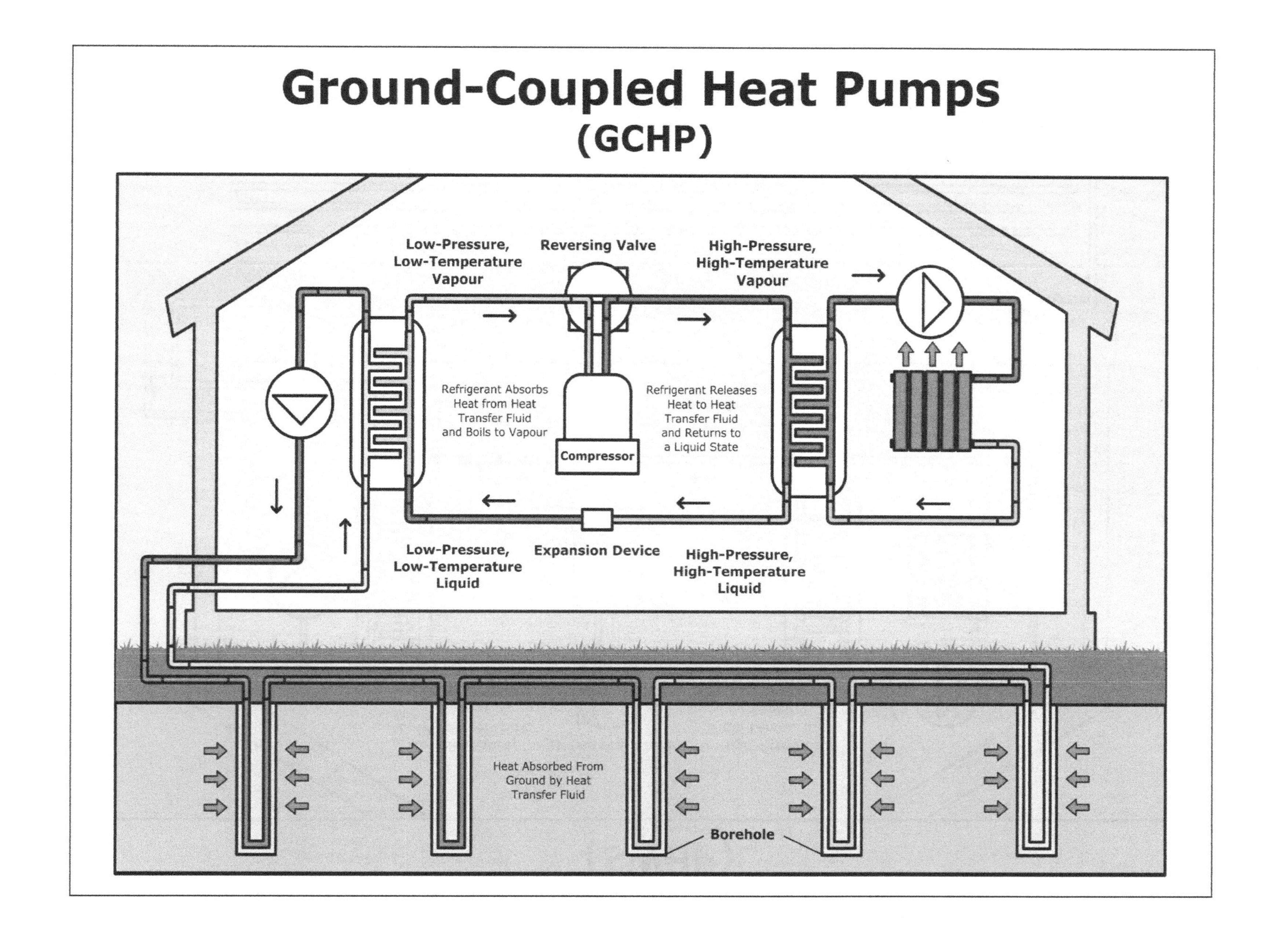

Ground-Coupled Heat Pumps
(GCHP)
Low-Pressure,
Low-Temperature
Vapour
Reversing Valve
High-Pressure,
High-Temperature
Vapour
Refrigerant Absorbs
Heat from Heat
Transfer Fluid
and Boils to Vapour
Compressor
Refrigerant Releases
Heat to Heat
Transfer Fluid
and Returns to
a Liquid State
Low-Pressure,
Low-Temperature
Liquid
Expansion Device
High-Pressure,
High-Temperature
Liquid
Heat Absorbed From
Ground by Heat
Transfer Fluid
Borehole

Hybrid Heating Systems

A hybrid heating system uses an intelligent system control (hybrid manager) to switch from the heat pump heating to the conventional heating system depending on the outside temperature and the efficiency of the system. When the heat pump operates at, say, low efficiency, the control system switches to the gas boiler. This ensures optimum efficiency and comfort all year round.

There are several applications for hybrid heating:

heat pump and a conventional gas heating system

This is another type of heating system as additional heating in case the heat pump cannot cope with the heating load (provide the required amount of heat).

Used in an old building where gas boilers and radiators are already installed. For the effective use of radiators, we need at least 140 °F (60 degrees Celsius), ideally a bit more – 167 °F (75 degrees Celsius). The heat pump can heat circulating water (which circulating through radiators) to 104 °F (40 degrees Celsius), in order to achieve another 86 °F (30 degrees Celsius) gas heaters are used.

heat pump and solar thermal system

In solar thermal systems, water is heated by the sun and then stored in a buffer tank (details on next pages). This heated water is then used to help the heat pump keep the water warm (e.g. in a floor heating system), reducing the running time of the heat pump. To make it even better, we can use a photovoltaic system, which provides additional savings on electricity bills.

There are also high-temperature heat pumps that can raise the operating temperature to 176°F (80°C). This is achieved by connecting two heat pumps in series. However, this results in high energy consumption due to the use of two compressors. The low efficiency is the reason why it is rarely used.

Buffer tank

In most cases, the buffer tank is a well-insulated, often cylindrical container that stores heated water or other heat transfer fluid in a heat pump system.

The buffer tank can have several functions:

Reduction of short-time operation. Reducing the frequency with which the heat pump switches on and off. Short cycling occurs when the system is too large, or the heating load is low. Short cycles cause the compressor to turn on and off too often, causing wear and early failure. By adding a buffer tank to the system, the compressor will run longer and rest longer because the heat stored in the buffer tank will keep the heated water in the system for longer, resulting in increased performance and longer service life.

Stabilization of the temperature. The buffer tank helps stabilize the temperature of the water circulating through the heat pump system. It acts as a heat accumulator, absorbing excess heat during periods of low heat demand and releasing it during periods of high heat demand. This helps maintain a constant temperature in the heating circuit and increases the overall efficiency of the system.

Reserve heat storage. Heated water is used as a reserve in case of failure or interruption of the heat pump system. This increases the reliability of the system.

Integration into the system. Used to integrate different heating systems into one. For example, the combination of solar thermal and heat pump or gas boiler, as already mentioned in the hybrid heating systems. The buffer tank stores and distributes the heat generated by each heating method in the system, maximizing the efficiency of the system.

Legionella

Legionella is bacteria that can multiply at ambient temperatures of 20 to 50 °C (68-122 ºF) of standing water, and a warm water tank (buffer tank) is a place where they can multiply rapidly. This becomes problematic when Legionella bacteria begin to grow in drinking water, which can cause disease. At temperatures above 60 °C (140 ºF), these bacteria die. People can become infected by inhaling water droplets containing bacteria, e.g. from showering or other sources that produce aerosols.

The heat pumps are equipped with an automatic legionella electric circuit, which disinfects the water by periodically raising the water temperature above 60 °C.

CHAPTER 4: PLANNING AND INSTALLATION OF THE SYSTEM

Design of the heat pumps

You should not design the heat pump capacity yourself but usually have it calculated by an energy consultant or the HVAC installer. If heating specialists do not offer a hea load calculation, you must ask an energy consultant.

The energy consultant can make much more detailed estimates and take into account the shape of the building. For example, a long, narrow house has more exterior walls than a square house, resulting in greater heat loss than a square house with fewer exterior walls.

The heating demand increases with decreasing outdoor temperature and the heat output of the heat pumps decreases with decreasing outdoor temperature (in case of air sourced systems). The outdoor temperature at which the heat demand and the heat output of the heat pump are equal is called the **bivalence point.**

If we want the heat pump to work at 5 °F (-15 degrees Celsius) without additional heating, we need to design the heat pump output to match the heat demand of the building at this outdoor temperature of 5 °F (-15 degrees Celsius).

For example, if the heat demand of the building at 5 °F is 15 kW of heat, we need to make sure that the heat pump can provide 15 kW of heat at this temperature.

If the designed heat pump will often turn off and on (in most cases when temperature outside isn't very low) – it means that it is oversized, which shortens the life of the heat pump. In this case, we need to use a large buffer tank, instead of installing a large heat pump, we need to find a middle ground and set the bivalence point a few degrees above the standard outdoor temperature.

Heat pumps must be serviced at least once a year (usually before the heating season - the beginning of autumn). Maintenance includes checking the refrigerant lines, inspecting and replacing filters if necessary, and checking the performance of the HVAC system. During maintenance, the most important thing is to check the refrigerant level and make sure it is not leaking. Next, we should check if the circulating water circuit isn't leaking and if the pressure in the system is still as high as it was in the last heating season (if the heating medium for radiators and underfloor heating is water).

If the compressor fails, the entire heat pump (indoor or outdoor unit) must be replaced in most cases. This is associated with high costs. However, some models allow you to replace the compressor, which is not so expensive. Pay attention to this point when choosing and buying a heat pump.

Summary: If the heat pump is too small, the compressor will run for too long, which will shorten its service life. If the heat pump is too large, it will turn on and off too often, which will also shorten the compressor's service life. The heat pump should turn on and off a maximum of every 20 minutes.

Choose an HVAC system with an inverter compressor as it has a longer lifespan than normal inverters. Install a buffer tank to reduce the number of times the compressor is turned on and off.

Installation of a HVAC System

Required Steps in the Installation of an HVAC System by a Professional Installer

1. **Assessment**. The professional installer should determine the heating and cooling requirements and select a suitable type of HVAC system for your home. The choice should be based on the size of the building, the level of insulation, and the available heat sources (availability of land and the ability to drill wells and so on).
2. **System sizing.** The installer should design the system for your conditions, determine the size and capacity of the heat pump, and select the most appropriate locations for the system components such as the outdoor unit and the indoor unit.
3. **Installation of the indoor unit.** For this purpose, usually, a technical room or utility room in the building is used. The installer installs the indoor unit (furnace) and connects it to the existing heating system of the house, usually radiators, floor heating, or air ducts if it is air heating, and then installs the electrical and control components to control the operation of the system.
4. **Installation of the outdoor unit or/and carrying out other work.** Mounting the outdoor unit on the building wall or nearby on a concrete slab. Digging trenches or building wells for laying heat pump pipes if we have a geothermal heating system.
5. **Connecting everything together** (electrical wiring and piping).
6. **Commissioning the system and testing its operation.** The installer puts the system into operation and checks all important working parameters, such as pressure in the system, refrigerant level, and operating temperatures. Heating and cooling capacity.
7. **Customer support and operational training.** The installer must explain how to maintain the HVAC system and provide ongoing support and services.

DIY for Beginners

The installation costs can range from $5.000 to $20.000. Installation of an HVAC system is a complex procedure that should be performed only by a professional. It should be done if a person has the necessary knowledge, training, and qualifications.

Most manufacturers require installation by a professional to maintain the warranty on their product. In the event of a fire in a boiler room or elsewhere, insurance companies may refuse to compensate for losses. The reasoning is that the system was installed by a non-professional, even if the cause was a defect in the components of the HVAC system (outdoor or indoor unit).

Errors during installation can lead to a reduction in system efficiency, resulting in higher operating costs and increased wear on system components. In addition, installation errors can cause safety hazards such as short circuits and fires, as well as liquid or gas leaks. So, obviously, it is better to entrust this to professionals.

To save money, we can do some preparatory work, such as pouring a concrete foundation for the outdoor unit of the system and dismantling and removing the old system in the house (gas or oil).

Do-it-yourself can be useful only if you want to install a simple air-to-air or air-to-water split system. And currently on the market is available many DIY-friendly AC or heat pump installations. *But first, you need to make sure you can do this legally.* If you decide to do the installation and/or wiring of the HVAC system yourself, be sure to document everything (take photos and videos of the work). This will help you prove the quality of the work performed when selling your house or in case of any problems.

DIY HVAC System Installation (AC/Heat Pump)

Many manufacturers advertise that some of their systems are suitable for do-it-yourself installation and provide detailed instructions for doing so.

Ductless mini-split systems, for example, are often considered DIY-friendly. They have outdoor and indoor units and require only a small hole in the wall to connect cables and pipes. If you do it yourself, it is very important that you carefully follow all the instructions.

Mini split heat pump system (ductless)

Tools and materials you need to prepare for doing an installation:

- Indoor and outdoor HVAC units
- Ductwork materials (if needed)
- Drill and drill bits
- Insulated screwdrivers set
- Wire strippers

- Level
- Tape measure
- Pipe cutter
- Pipe wrench
- Hammer

The general process of installing a ductless mini-split system consists of several stages:

1. Choosing a suitable HVAC system for your needs.
2. Select and prepare a location for indoor and outdoor units and prepare all tools and materials for work.
3. Drill a hole in the wall for connecting the indoor and outdoor units with cables and refrigerant lines.
4. Attach the mounting brackets for indoor and outdoor units. Can be easily purchased online if not provided by manufacturer. Secure the units and make sure they are level. Make sure that the outdoor unit has enough space for airflow. Clear all debris.

   If on the ground: Lay down a solid base, this can be a concrete slab or poured flat concrete surface on which to place the unit.
5. Connect the refrigerant lines to the indoor and outdoor units. Run the copper tubes through the wall (in most cases, the tubes are ready-made from the manufacturer). After that, we can connect the ends to the valves on the outdoor and indoor units and secure them with a wrench. More details is provided by manufacturer.
6. Now we can install ductwork (if needed by design). Cut ductwork materials so that we can connect them in the building. When ductwork is installed, you can use mas-

tic of foil-backed tape for insulation and sealing ductwork. Attach ductwork to the inside unit (handler). In the case of ductless systems, this step is not needed.

7. Connect the electrical wires between the indoor and outdoor units, and then connect the outdoor unit to the power supply. Detailed instruction is provided by manufacturer.
8. Thermostat installation. Choose a place for it in the building (best in the center of the building for optimal home temperature readings) and install it at eye height for convenience. Read and follow instructions provided by the manufacturer.
9. Charging the system with refrigerant. To perform this step, it is better to contact an HVAC specialist. This requires expensive professional tools and qualifications (license).
10. Finally, we need to test the work of the system to make sure that everything is working properly.

The test procedure and the data of the working tests for each model of DIY Heat Pump/AC are usually described in a manual for installation.

It is not possible to write very detailed installation instructions because each HVAC system has a bit different installation process. Therefore, there are separate instructions for each model and not one universal.

For beginners, it is a great start.

The most important things to keep in mind when installing an HVAC system in an old building

First, you should consider some important factors:

1. **Insulation** - many old buildings are not properly insulated.

2. **Upgrade or replace the electrical system** (cables and fault breakers) in the house. A heat pump can have a high-power demand and it is better to ensure that its use is safe.
3. **Evaluation of the available space for a heat pump system.** In an old building, space is often very limited, which complicates the installation of the system.
4. **Is it possible to install underfloor heating?** If not, is there enough space for larger radiators? Is there space for low-temperature radiators?

   Low-temperature radiators have a much larger heat transfer area, which allows for efficient heat transfer from the radiator to the air in a room with a low circulating water temperature (about 40 degrees Celsius). Often used for heat pumps.

5. **Is it a historic building?** For some old buildings, there are regulations and restrictions. You need to make sure that you can make all the necessary installations. If in doubt, you should contact your local authorities.

CHAPTER 5: TROUBLESHOOTING AND REPAIR OF THE HVAC SYSTEM

Required Tools for Troubleshooting and Repair

Safety tools you must have:

- **Rubber gloves** – you will work with high voltage - 240 V. And everything above 30 V can cause electrical shock. 110 V and 240 V can be deadly if you accidentally touch any metal part that is live.
- **Insulated screwdrivers** (helps to avoid contact with electricity)
- **Safety glasses to protect your eyes.** If you accidentally touch live parts of the system and cause a short circuit. Some of the metal can boil in a split second, and this metal can get in your eyes or on your hand. It would also be ideal to use the jacket to protect your hands.

Safety equipment that might be helpful to have:

- **Pressure or relief valves** (to keep pressure in the system at safe levels)
- **Emergency shutdown control** (to stop the system in emergency situations)
- **Fire protection devices** (smoke detectors and so on)

Basic tools you need:

- **Needle-nose and tongue-and-groove pliers**
- **Screwdrivers with insulation (electrical)**
- **Tape wrenches**

- **Multimeter/Voltmeter**
- **Pipe wrenches**
- **Torque wrench**
- **Manifold gauges for pressure measurements**
- **Wire stripers for removal of wire insulation**

Other tools for experienced people:

- **Vacuum pump**
- **Electronic leak detectors**
- **Refrigerant recovery machine (for safe removal of refrigerant)**

HVAC tools you may need:

1. **Refrigerant leak detector** (allows you to measure small refrigerant leaks in fittings and pipes that are not visible)
2. **Manifold gauge set** (in case of work with refrigerant lines)
3. **Anemometer** (to measure air flow rate)
4. **Infrared thermometer** (to hot and cold spots in the system and in the building)
5. **Combustion analyzer** (to test heat exchanger in a furnace - air to fuel ratio and availability of harmful gases)
6. **Vibration analyzer** (for detecting patterns in vibrations during operation and finding a problem)
7. **Electronic or digital micron gauge** (allows you to display very accurate pressure data)

How to Do Troubleshooting of the HVAC System

We should start from inside the house

1. **Check the thermostat if it is working and if the temperature is set for the right temperature.** Check that the thermostat is on and receiving power. If the thermostat has built-in batteries than check if they are charged. Just try new batteries, they cost only a few dollars. If everything is good here, we can move to the next step.
2. **Check all fuses and circuit breakers in the system.** Check main circuit breakers that power the HVAC system. Inside the house where the system (evaporator) is installed, there is usually a control board with fuses. Typically, these are automotive fuses, such as the ones shown in the figure below.

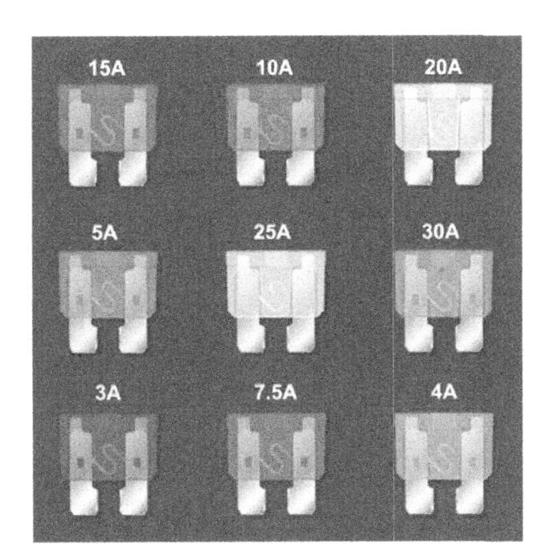

Or there could be some in-line fuses

If some of these fuses blow up, you will see that the metal wire inside it is a melt-down. To replace this fuse is super simple – just look at the fuse rating and buy the same one in any car repair store (usually these are fuses from 2 to 10 Amps). Some of these fuses are located on the **furnace control board** (more details later). Then just pull out the old fuse and insert the new one. But before doing this you have to turn off the circuit breaker – to be safe. Costs a few dollars, but an HVAC contractor will probably charge you a couple hundreds of dollars. Robbery!

3. **Next, check the control board itself.** If you notice that it is burned, there are signs of darkening, then you will most likely need to replace it. You will need to look online for the same control board and ask an HVAC professional or an experienced friend to replace it. Sometimes it's very easy to replace, some boards have connectors, and we just plug and unplug the cables. This problem can be caused by sudden power surges. Also, while you're here, it's a good idea to check the wires and the contacts on the wires to see if they look good and are well connected. But be careful not to pull too hard to avoid damaging them.
4. **Check if visually the system doesn't have any problems.** Check the air filters in the furniture of the house if it clean and allows good airflow (replace them if needed). Check the vent and ductwork for blockages.

   Very important to see if there are no leakages of refrigerant from the evaporator coil inside the house. This can lead to poor system performance or stop the system from working. If you can see the clear signs that it is leaking from the coil, it means that we need to replace it and refill the refrigerant in the system. This is very expensive for the homeowner. In this case, the cold or hot air (in the case of a heat pump) will not be blown from the fan.
5. **Check if the evaporator coil inside the house is clean of dirt and debris.** Again, this leads to poor heat transfer and, as a result, the system does not cool or

heat the air as required. The evaporator coil will need to be cleaned (ideally at least once a year), but usually no one doing this until a problem occur (no cooling or heating).

6. **Check the blower motor for proper operation and the amount of current (Amps draw) supplied to it.** If you hear it running, there is no problem. If you do not hear the fan motor running, it means that it has either failed or burned up or the fan is stuck, and the motor cannot move it. In this case, you can try to spin the fan to see if it starts and to understand if it is spinning well. But be very careful when doing this. There are many reasons why a fan might not work, and we will discuss them later in this book.

   Also, the blower might work not sufficiently, and you don't get enough air circulating the house. Use an ampere meter to measure the amount of amps that going to the blower and compare it to the amp draw that is provided by the manufacturer of the blower. This number can be easily found online or on the fan motor label. If it's a number lower than provided by the manufacturer – it means that we need to replace it or give it to a professional for repair. In most cases, it is replaced. Another reason is a faulty capacitor. There are many reasons why a fan might not work, and we will discuss them later in this book.

7. **Additionally check the evaporator's condensate drain line.** If it is blocked, a lot of condensation will leak out around it. This also affects the air quality as the condensation will start to evaporate. You also have a risk of mold growth due to high humidity and dampness.

Going outside the house to the outdoor unit

8. **Checking if there is no leaking from refrigerant lines and condenser coil inside and around the unit.** If noticed some, should be made pressure check and

comparison to recommendations of the manufacturer. We have to hire a licensed HVAC technician.

9. **Checking if everything is tight and right** – the protective panels on the outdoor unit and if the unit is firmly on the ground. If the bolts are not tightened, you will hear strange noises from the outdoor unit (vibrations).
   Additionally, the leveling of the outdoor unit is important - if it is not properly leveled, it can also cause vibrations.
10. **Check if the condenser coil is clean.** If it isn't, clean it from the inside out of the unit. For doing this you will switch off all power to the outside unit by using the power switch and take off the panel to get access to the inside of the unit. With this method, dirt will be removed outside the unit.
    But it is also ok, to clean it from the outside as well. You just must understand that one day you will still have to clean the inside of the unit.
11. **Check if the fan is working when the outdoor unit is on.** If not, check to see if the fan is turning with a screwdriver (be careful). Rotate it in the direction it should be rotating (the blades should be scooping air outward). If you cannot spin it up, it means that the motor fan should be replaced.

Sounds/Noises the HVAC System Can Produce

1. **Buzzing outside unit.** Usually, it is a contactor. Tap on it with an insulated screwdriver, if the sound changes, then the problem is in it. If the noise doesn't go away, then you need to replace it.
2. **Screeching fan motor.** Metal-to-metal sound (grinding noise). The fan motor needs to be replaced or you can try lubricating the fan motor (this can greatly reduce or stop the noise for a long time). Fan motors may have oil plugs, but in most cases they do not, in which case we need to lubricate the motor shaft from above and from there the oil will be sucked into the

motor (bearings). This usually requires removing the fan motor from the outdoor unit.

3. **Rattling noise** (vibrating metal-to-metal sound). This usually occurs when the screws on the outdoor unit become loose and the outdoor unit panels start to vibrate, causing this sound. All you need is to tighten the screws and the sound will go away. You can put something under the screws (a washer) or put a rack or brick on them, and the sound will disappear. But make sure that air can easily flow out of the outdoor unit.
4. **Loud compressor when the unit is running** (grinding, screeching, clanking - depending on the compressor). Probably will require the replacement of the unit or compressor. A professional should look at it.
5. **Bubbling sound.** Sometimes it can be caused by refrigerant lines. Air bubbles are traveling the refrigerant lines. Better to not do anything about it, because fixing this sound is very expensive.

   Or it can be a drainpipe or port below the evaporating coil in the house that is partially plugged. Simply clean the condensate drainage pipe with a vacuum cleaner to suck everything out of the pipe.
6. **Brocken/cracked fan blade.** Check that the blade is in good condition. If not, you need to replace the fan blade. Sometimes it's hard to notice that it's cracked, but when the blades rotate, it starts to make noise and crack further.
7. **Rapid clicking sound in the outdoor unit while working.** Usually, this might be a loose wire on the thermostat. Carefully check all wires in the thermostat if they are well connected. Discharged batteries of a thermostat can also cause this sometimes.
8. **Whistling noise inside the house in the furnace area.** Most likely the evaporator coil is frozen (the outside refrigerant line can al-so be frozen),

or the sound can be caused by a very dirty air filters, or blocked air return drills can also cause this.

9. **Hissing noise.** In this case better to ask an HVAC technician to check it. Most likely it is a sign of a refrigerant leak somewhere in the line.
   Also hissing can come out of the outdoor unit (compressor) which releases pressure (a very loud sound). Turn off the HVAC system and have it serviced by a technician - it could be dangerous. This may be due to a faulty fan motor or it's capacitor in the outdoor unit (the condenser coil is not cooled, which leads to an increase in temperature and pressure in the system). If the fan does not work, you need to find out why and try to fix it. Clean the condenser coil - if it is clogged, heat transfer will be poor, and the coil will start to heat up. ***If the fan and condenser coil are in good condition, you should have the compressor checked by a specialist.***
10. **Pulsating noise.** This sound can come from refrigerant lines. In places where they pass through walls, the refrigerant lines can start to vibrate and cause this sound. This occurs when the insulation of refrigerant lines in the walls wears away over time. Check these areas inside and outside the building and add insulation (rubber, foam) to prevent direct contact with the walls.
11. **Noise inside the house** – just trace it down to find from where it is coming. Usually spotting and fixing it is easy.

The Most Common Problems with HVAC System

1. **Bad capacitor. The sign:** the fan inside the house will run and blow room-temperature air or even hotter. But the outdoor unit will not work at all. This may mean that the capacitor has completely failed. In this case, you may hear the compressor trying to work ("buzzing") every 20-30 seconds. In most cases, it is a

capacitor. A new capacitor can cost from $10 to $30, and the replacement process is not complicated.

Capacitors often fail, especially if it is hot outside for a long time (several days or a week), there is a high probability that the capacitor will fail. For reference, a power surge can also damage a capacitor.

Another situation is possible when the fan is blowing in the house and the fan is blowing outside, but the compressor cannot turn on either (you can see and hear that only the fan is working (+ buzzing from time to time)). The compressor gives a little bit rumbling sound when working. This may also mean that the capacitor needs to be replaced.

The third situation that can be caused by a capacitor is that fan of the outdoor unit isn't working. The compressor is running, but the fan isn't. You may even see smoke coming out of the outdoor unit because the condenser coil isn't cooling, resulting in very high temperatures.

It can also mean that the fan in the outdoor unit has failed, if you try to spin it and it rotates, it means that the problem is most likely in the capacitor.

2. **The condensate drain line is clogged.** Everything is working great, but you can see the water on the floor around the air handler of the furnace. This problem can be solved easily. In this case, the fitting or condensate drainpipe is blocked. Use compressed air to push it out of the pipe or a vacuum cleaner.

   The reason for the blockage of the condensate drain is that the filter is rarely changed, which leads to a lot of dust getting into the condensate and clogging the pipe. Mold can also block the flow of condensation through the pipe.

3. **Low level of refrigerant in the system.** In this case, everything is working (fan inside the house, and fan, compressor in outdoor unit) but you either do not get cold air or the room temperature cannot drop to the temperature set on the thermostat temperature (not enough cooling). First, we need to check if the condenser

coil and evaporator coil are clean and don't have obstacles to heat exchange (if not we should clean them). The second thing to check is air filters (replace them). If after doing all this, you didn't see a significant difference in cooling then the last thing to do is to check the return and supplying air temperature in the furnace (supply can be measured on the closest vent and return temperature where the filter is). The temperature difference should be between 15- and 20-degrees Fahrenheit on a thermometer for well-functioning systems. If it is less than this number, then most likely there is not enough refrigerant in the system. In most cases, it is best to contact a specialist to have the system recharged (as we are not licensed to do so).

4. **Control board.** For example, the outdoor unit is working fine but the fan inside the building isn't working which leads to the freezing of the evaporator coil (but it is not always because of the control board). Sometimes the system will start to behave strangely – everything can start running at the same time. The fans can run continuously without stopping.
5. **The fan of the outdoor unit has failed.** If you can't spin the fan or it barely moves, it needs to be replaced. If the fan blades rotate well, then it is most likely the capacitor.

   Sometimes at the beginning of the season, the fan can be a bit stuck – so you need to move it slightly to get it to start rotating. Be careful not to hurt yourself. Use a long screwdriver and turn (push) the fan blades - the blades should push air out of the outdoor unit. Often the fan starts working as usual after doing this.

Replacing or Cleaning the Air Filters

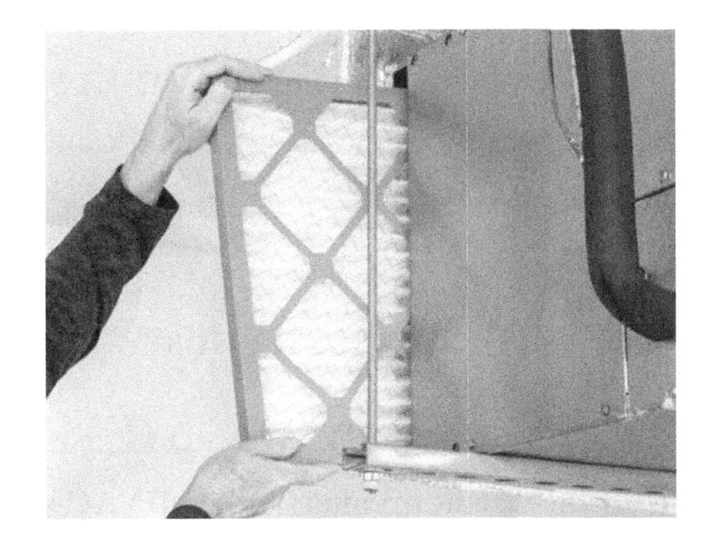

1. Air filters should be changed every month or less if it is a **simple fiberglass filter.** This type of filter is recommended by most HVAC manufacturers because it offers the least resistance to airflow. However, the disadvantage of these filters is that they allow a certain amount of dust to pass through, and over time, the evaporator coil can become clogged, causing it to freeze. They are thin and usually green or blue in color.
2. **If you have washable electrostatic filters.** They should be cleaned every month (30 days). They have plastic frames. The downside of this type of filter is that you will need to clean them.
3. **Plated filters must be replaced from 1 to 3 months (90 days).** But it is better to do it more often (monthly), especially if you use the air conditioner or heating frequently. When you check the air filter after 1 month of use, it may appear to be clean, but the airflow will be restricted. You will probably notice that it starts whistling and the fan is running much more intensely than usual (this can lead to overheating and burnout). I always recommend replacing the filters every month or more often if you use your air conditioner constantly and have children and/or

pets. This type of filter is recommended by professionals for most HVAC systems. They are cheap and available in many stores.

4. **High-efficiency filters should be replaced every six months.** These are relatively expensive filters and restrict the airflow significantly in the system. They are thick and have a metal frame. This is not recommended for residential buildings.

How to do this

1. **First, we need to turn off the HVAC system – use the switch near the furnace or turn off the circuit breaker for the HVAC system.** We do this to prevent the fan motor from running (stopping the airflow) while we change the filter.
2. **If you don't know it, then check where is located the air filters.** Almost always it is located before the blower motor in the return ductwork of the furnace. If you can't find it, ask someone or look in your HVAC manual. Sometimes it can be located inside the furnace next to the air blower, you will need to take the doors of the furnace off. Rarely, but in some systems, filters can be located in the wall or ceiling. The air filter slot will have a cover that will need to be removed to get the filter out, if there is no cover you will need to tape it with duct tape. Also, you can find magnetic strips online that might be used for this purpose.
3. **Check the manufacturer of the air filter** that is already installed and its size so that you can buy a new one in a store (amazon or any other local store). Another important thing to pay attention to when changing the air filters is its MPR (microparticle performance ratio). The higher the MPR is the more dust it catches up, but the downside of this is restricted airflow in the system.

   Typically, air filters with an MPR of 300 to 600 max are the best for residential buildings. In this case, we have a good airflow rate and filtering capacity.

Some filters have a MERV (Minimum Efficiency Reporting Value) rating, which should be between 3 and 5 for good airflow in the system.

4. **Carefully pull out the old air filter from the airflow line (air return).** On a new air filter that you bought should be a pointer (arrow) for airflow. The arrow should point in the direction of airflow (pointing to the air blower which sucks the air in).

   If you notice that you don't have an air filter in the return duct, you can take measurements (width, length, and thickness) with a tape measure. All HVAC systems must have them (probably someone forgot to install it back).
   In this case, you will need to take measurements of the duct air filter slot (width, length, and thickness of the slot). To do this, you can use a tape measure or a regular ruler.

5. **Insert the new or cleaned air filter back into the slot.** As I said, make sure everything is done correctly and the filter's arrow is pointing in the right direction (to the air blower). Close the air filter slot with the cover.
   Tip 1: Write down on the air filter the date when it is installed in the furnace so that you could always know how long the filter is doing its job.
   Tip 2: An arrow can be drawn on the return air duct to match the arrow on the air filter. This will help to prevent the filter from being installed incorrectly.
   Tip 3: Write the dimensions of the filter on the air duct near the arrow so that you can always read them when needed.
6. **Now, when everything is ready, we can run the system again.**

Replacing Bad/Failed Capacitor (in outdoor unit)

First, you should know what type of capacitors your HVAC system has. There are standard run capacitors or "dual run capacitors". Dual-run capacitors mean that two normal capacitors are combined in one (usually used for running compressor and fan motor in outdoor unit). Dual capacitors have three terminals instead of just two.

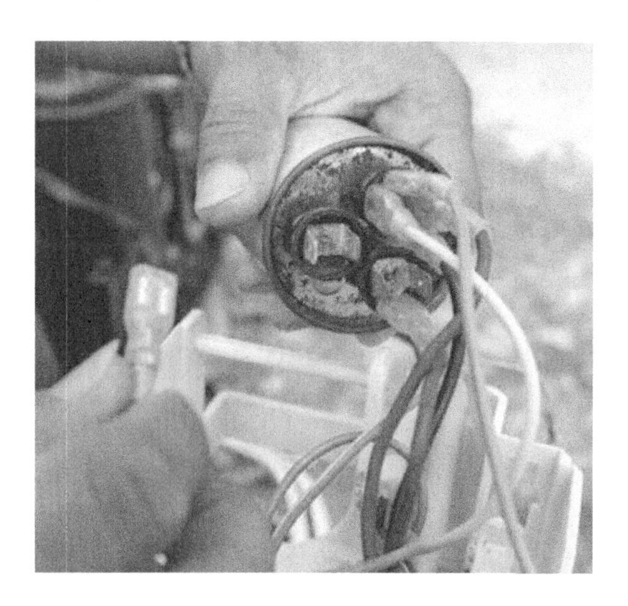

You can visually see that the capacitor is bad - it usually starts to deflate.

Terminal labels for a capacitor:

HERM – "hermetically sealed compressor" for the compressor.

FAN – as you might guess, stands for a fan motor.

C – mean common/neutral.

A standard capacitor has two terminals - C and FAN or HERM if it is designed for a compressor. **But dual – all three.**

How to:

1. **First thing to do is to use the disconnect switch to turn off the power.** It is usually located on the side of the house (plugged disconnect with a handle that you will have to pull off). But sometimes it can be a switch or a breaker. Also, turn off your thermostat to the off position so that the contactor does not get 24 V of power as well.
2. **Remove the panel**. All electrical components should be located where the electrical cables (whip) are connected to the outdoor unit. To access them, remove the panel by loosening a few screws.
3. **Find a cylindric run capacitor. Read the label on it and buy a new one with the same characteristics.** If you can't see the label information on capacitor, check the compressor label and fan label (usually there is mentioned ratings for capacitor – you can read **CAP** and then rating in microfarads and AC voltage). If you can't read the labels on the components, search online for the specifications of your HVAC model.

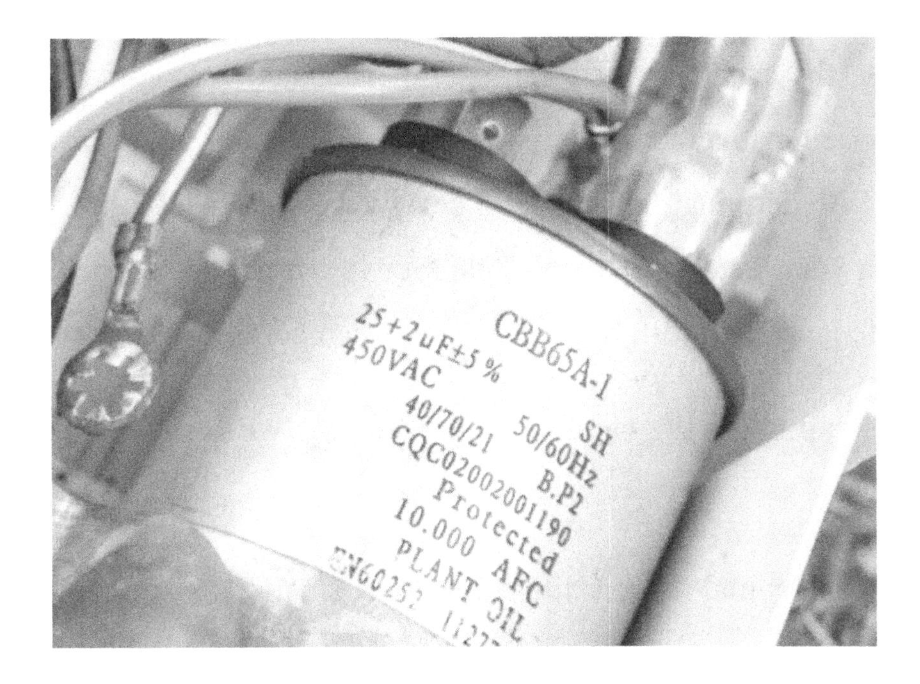

Before removing the capacitor from the outdoor unit, take a photo of the wire connection (make sure you can see all the markings on the capacitor and the color of the connected wires) and remember that it may be charged, which could result in an electric shock. Alternatively, use a voltmeter or voltage meter pen to check that the wires are not live.

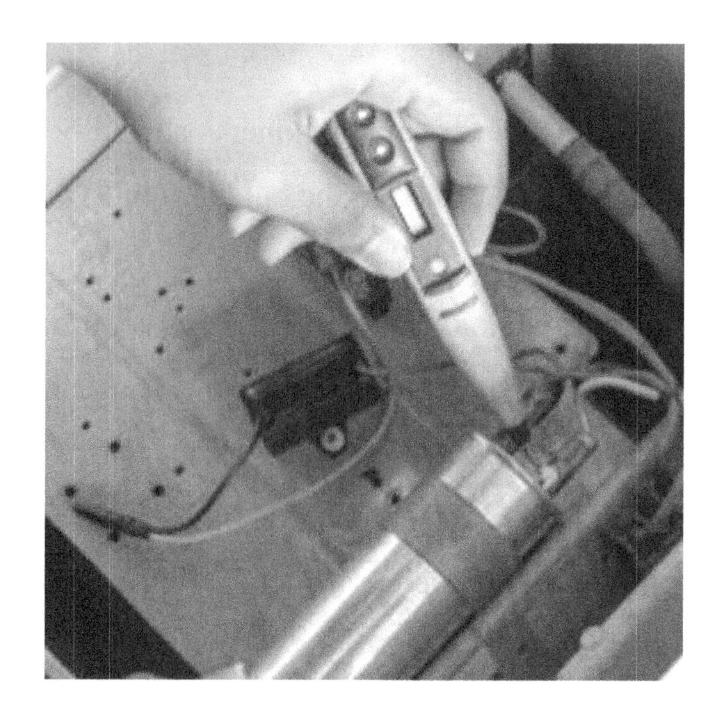

To discharge the capacitor, you can use an insulated screwdriver and connect the terminals of capacitor together in pairs. First C and HERM and hold it for 5 seconds, then C and the FAN terminal for another 5 seconds as well. Make sure you do not touch metal parts with your hands. After doing this for few times the capacitor should be discharged to a safe level. But still, be careful and remember to wear rubber gloves.

The metal strap holds the capacitor with one or two screws that we will need to unscrew.

The new capacitor may look a little different (smaller or larger) than the old one, which is normal - just make sure the ratings are the same (mF and AC Voltage (VAC)). The most important number is microfarad (μF) number, which must be the same as on old capacitor. The voltage may be higher, but not lower than the voltage indicated on the old capacitor.

25 + 2 mF (+/- 5%), 450 Volts dual-run capacitor

25 mF – is rating for compressor

2 mF – is rating for fan motor

+/- **5%** – are the limits within which these ratings can be higher or lower when you measure the mF rating on it. **If you measure more than 5% than it means that capacitor is bad.**

For example, if your multimeter reads **22 µF or 28 µF** between the common terminal (labeled C) and the compressor terminal (labeled HERM), which is more than 5% (in our case, more than 10% difference from 25 µF), then the capacitor is bad. Make sure that your multimeter **can measure µF for capacitors and select the µF measurement** on it.

To measure smaller rating for fan motor you will need to measure **µF** between common terminal (C) and fan motor terminal (FAN) and make sure it **isn't bigger or smaller than 5% of 2 µF.**

How do I calculate a rating that will be 5% higher?

25 µF x 1.05 (5%) = **26.25 µF** (the maximum rating) which is 1.25 µF bigger than 25 µF

1.25 µF is 5% of 25 µF (5% = 1.25 µF)

If the multimeter reads more than **26.25 µF**, the capacitor is defective.

How to calculate the smaller number (-5%)?

25 µF – 1.25 µF (5%) = 23,75 µF (minimum µF rating)

If the multimeter reads less than **23.75 µF**, the capacitor is defective.

After the capacitor is discharged, you can take a picture of the wires connected to the old capacitor. Or you can disconnect one wire at a time and connect it to the new capacitor one by one – *this method is very good for beginners.*

Look at the connectors on the wires to see if they are in good condition and not corroded. If you see that they are defective or broken, strip the wire and install a new connector on the wire.

4. **Put a capacitor in the metal strap.** If the new capacitor is smaller than the old one, you can squeeze the strap with pliers in most cases.

5. **Check the operation of the outdoor unit** by setting the thermostat to cooling or heating (in case of heat pump). Listen to the compressor in the outdoor unit and whether cold/hot air is being delivered to the house.
6. **Screw the electrical panel** of the outdoor unit back on.

It is always a good idea to have a few spare capacitors for your HVAC system, especially during the summer when they fail often.

Replacing The Contactor in the Outdoor Unit

This rarely is the reason your HVAC system isn't working, but sometimes the problem might be in the contactor.

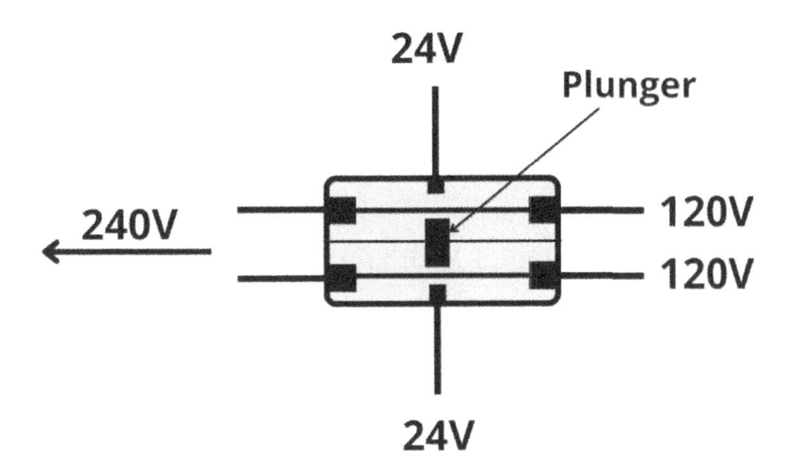

Reasons Why the Outdoor Unit Isn't Starting - Because of Contactor

1. **The contactor is very old and gets loose** – the plunger in the contactor can't go all the way down and make good contact for 240 V power. As a result, the outdoor unit does not start, or starts but does not operate properly. In the contactor, the plunger starts to heat up and can cause a fire, or at some point it is welded to the terminals and the compressor with the fan starts to run continuously without shutting down.
2. **The electrical coil in the contactor gets weak or fails.** This electric coil is powered and controlled by a thermostat (control board) which sends 24 V power to the coil. This coil creates a magnetic field that pools the plunger down to close the 240 V electric circuit which runs the compressor and fan. Sometimes the electrical conductors on the coil begin to weld together, and as a result, the coil stops working. If the coil is loose, the outdoor unit will switch on and off very frequent-

ly (e.g. several times per minute). The coil is not able to hold the plunger firmly in contact with the terminals.

3. **A bug or dirt has gotten under the plunger of the contactor**, which prevents the 240 V electrical contact from closing (the bug serves as an insulator). In this case, the outside unit just do not start working. To solve this issue, you need to blow compressed air under the contactor plunger to remove the bug or dust.

Contactors are very cheap at around $10, so replacing them can be a good idea even if you're not sure if the contactor is the problem.

Testing the Contactor

1. **Checking the electrical coil of the contactor.** On the switched-off HVAC system, including the thermostat (with using a circuit breaker and disconnect switch) unscrew the 24 V wires that are coming to the sides of the contactor (some old contactors can have two wires on one side).

   Set your multimeter to measuring resistance - Ohms (sign Ω) and connect two leads to the side terminals of the contactor. The multimeter should show between 12 and 27 ohms, if it is less than 12 or more than 30, it may mean that the coil is not working well, and the contactor will soon fail. If it shows the OL sign or zero, it means that the coil has failed, and the contactor must be replaced.

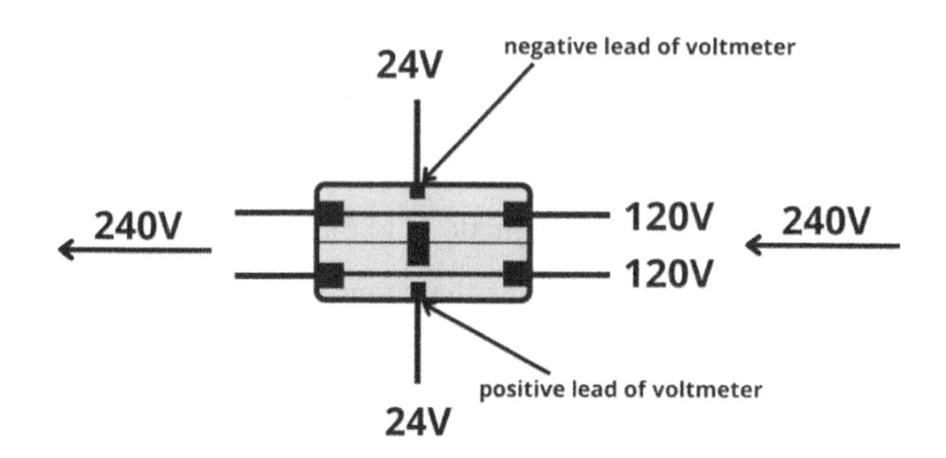

2. **Check the voltage at the contactor for the 24 V terminals when the thermostat is set to cooling and should be supplying power to the outdoor unit.** Be very careful and use personal safety equipment. Set your multimeter to voltage measuring and attach 2 leads to the side terminals of the contactor.

   Your meter should show from 24 to 30 volts. If it will show 0 volts – it means that the contactor doesn't get power. **The problem may be a blown fuse on the control board or a tripped condensate overflow switch.**

How to Replace

1. **First you turn off the main circuit breaker** in the house and pull off the disconnect that usually is near your outdoor unit.
2. **Then remove the electrical access panel on the outdoor unit.**
3. **Check if the thermostat is off** (to prevent 24 V from going to the contactor). If not, set it in off position.
4. **Check for a 240 V supply using a multimeter or voltmeter (see the image).**

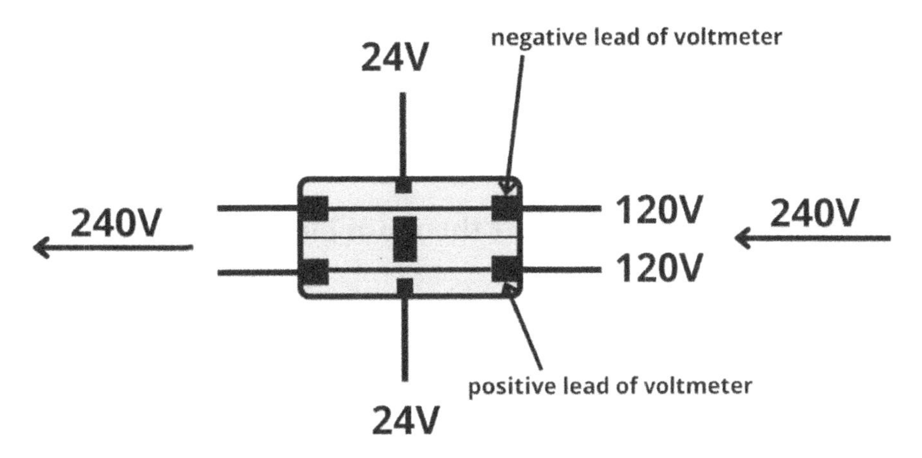

5. **Discharge the capacitor** using an insulated screwdriver (shorten (connect) the terminals – N to HERM and N to FAN).
6. **Check with a multimeter or voltmeter if there is any voltage present in the wires. Your multimeter should show around 0.**

7. **Check what kind of contactor you have installed** (single pole or double pole) and buy a new one, ideally the same one.
8. **Remove the bolts holding the contactor.** Take pictures of how wires are connected (to which terminal they go).
9. **Unscrew two wires on the sides of the contactor (these wires send 24 V power).**
10. **Unscrew the power supply wires (labeled on the contactor as L1) and screw them onto the new contactor.**
11. **Disconnect the remaining wires from the old contactor (L2) and connect them to the new one.**
12. **Once everything is ready and all wires are connected – you can secure the contactor in the unit with screws.**
13. **Put the electrical panel back on the outdoor unit and test** if the HVAC system is operating properly.

Replacing the Thermostat

1. **Buy a thermostat that you liked.** Modern thermostats can have Wi-Fi and a mobile app for managing home temperature.

2. **Turn off the HVAC system at the main electrical panel of the house.**
3. **Use a multimeter or a voltage pin** to check for voltage on the wires. It is an additional safety measure that needs to be followed.
4. **Take off the old thermostat and mounting plate**. Unscrew or unclip the thermostat and create a diagram of where each wire goes for later installation of the new thermostat. Take a picture of connected wires.

5. **Mount the new plate and level it.** Refer to the thermostat manual and diagram for installation instructions. This will be slightly different for each thermostat. Insert each wire into the terminal and tighten it with a screwdriver.
6. **Install batteries if needed and attach the new thermostat to the mounting plate.**
7. **Turn on the power for the HVAC system and check the thermostat work.** Set the desired temperature and other parameters for the system.
8. **For a few days – keep an eye on the HVAC system if it works correctly.**

Troubleshooting and Testing for Fan Motors

1. Check that the fan blades move easily.
2. Make sure that the contactor is supplied with 240 V power and that the plunger pulls in when the unit is starting. Check if the contactor is in good condition. Be very careful and wear safety glasses and gloves.

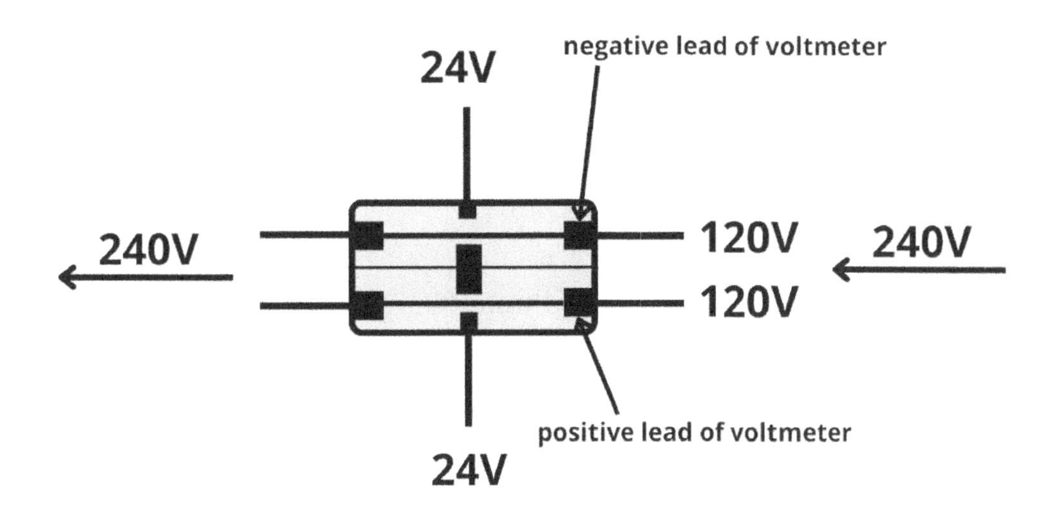

3. Check the capacitor to see if it is in good condition (does not deflate) and if the compressor starts. If capacitor is in good shape and compressor starts, then the fan motor is most likely defective.

How to Test the Fan Motor

1. Turn off the HVAC system (circuit breaker and thermostat).
2. Make sure that the wires aren't live by using a voltmeter or multimeter and find the wires that go from the fan motor (trace them).
3. Label them and take a picture of where they are connected.
4. Disconnect all fan wires (usually it is 3 wires).
5. Use a multimeter for checking resistance between wires (ohms). The multimeter should have the sign - Ω. We need to have 3 readings between all 3 wires.

The two smaller readings together should equal the largest number (a discrepancy of a few ohms is not a problem). If so, the fan is working properly. If not, the fan motor is defective and must be replaced.

Example:

Between black and purple wire - 60 Ohms

Between brown and purple wire - 65 Ohms

Between black and brown wire - 125 Ohms (60 + 65 = 125 Ohms)

If you get a much larger or smaller number than 125 Ohms (100 or 140 Ohm), this is a sign of a problem.

Sometimes, you may find continuity between some of the fan wires and the green (ground) wire that is connected to the outdoor unit casing. This is also a sign of a faulty fan. It should usually show OL (open line) on the multimeter or zeros.

How to Replace it

I don't recommend doing this for beginners because it's a complicated process and requires a lot of time and effort. But I will write a short description of the process so that you can understand it and do it if you want.

How to:

1. Switch off the power for the HVAC system and check if the wires aren't live by using voltage pin or multimeter.

2. Carefully take off the outdoor unit panel to which the fan motor with blades is connected. See the image above to understand how it should look when you took it off.

3. Cut off the wires that run from the fan motor to the electrical section of the condenser unit. The new fan motor will have its own wires.

4. If you buy a new fan motor with fan blades attached to it (it makes the process much easier). Just secure the new fan motor to the panel. Make sure the fan blades are facing the right side (pushing air out of the unit).

5. Take a picture of old wires that you already cut – to which terminals and component they are connected.

6. Put the panel with the new fan motor back on the outdoor unit and run the wires from the fan motor to the electrical section of the outdoor unit.

7. Remove the old (cut) wires and replace them with new ones (do this one at a time).

8. Check if the fan blades aren't too close to the wires going from the fan motor. If you see a risk of contact with the fan blades, try pulling the wires or securing them to the panel with zip ties.

9. Screw down all screws of the fan panel and electrical section panel.

10. Test if the outdoor unit is working and if the blades of the fan are spinning. Again – double-check if the air is coming out of the unit. Otherwise, the condenser coil of the AC unit will not be cooling well and cause overheating and other problems.

The Evaporator Coil of an Air Conditioner or a Heat Pump Freezes - Troubleshooting

1. Too low temperature on the thermostat. Most air conditioners are designed for a cooling temperature of 67-70 degrees Fahrenheit and if on the thermostat you put 61 then it will make your evaporator too cold, and condensate will start freezing on the evaporator coil and accumulate.

2. Low refrigerant level in the system. The pressure goes down in the system and the temperature as well goes down, which makes the evaporator colder than usual. When it is colder than needed the condensate on the coil starts freezing and builds up. In this case, we should ask a professional to add refrigerant to the system.

3. The metering device (expansion valve) is clogged, which leads to a decrease in pressure in the evaporator coil. The lower the pressure, the lower the temperature in the evaporator coil.

4. The evaporator coil gets plugged up by dust and debris, which lowers the heat exchange between air and the evaporator coil. As a result, the condensate begins to freeze. This requires cleaning of the evaporator coil.

You can use a duct cleaning company to do this or try to do this yourself. But be very careful and don't move the evaporator coil to prevent creating cracks and refrigerant leaks.

5. Dirty air filter that restricts airflow in the system. The evaporator coil inside the house (for conditioning mode) doesn't have enough air for absorbing heat from the house, and ice (condensate) starts to accumulate. The refrigerant line (thick line) will also start to freeze outside the house.

6. Air vents are closed. Too many vents in the house are closed which causes the same problem as with plugged filters (restricted airflow through evaporator). The solution to all this is to allow proper airflow by changing air filters and checking vents and air returns.

7. Blower motor not working. The evaporator does not receive enough air to absorb heat (the evaporator does not "heat up", does not absorb heat), and the condensate begins to freeze. The cause may be a control board that stops supplying power to the fan motor inside the house. But the fan motor can also fail.

Most of the issues mentioned above can be fixed by a beginner.

CHAPTER 6: TROUBLESHOOTING OF THE FURNACE AND MAINTENANCE OF THE HVAC SYSTEM

Furnace Troubleshooting and Repair

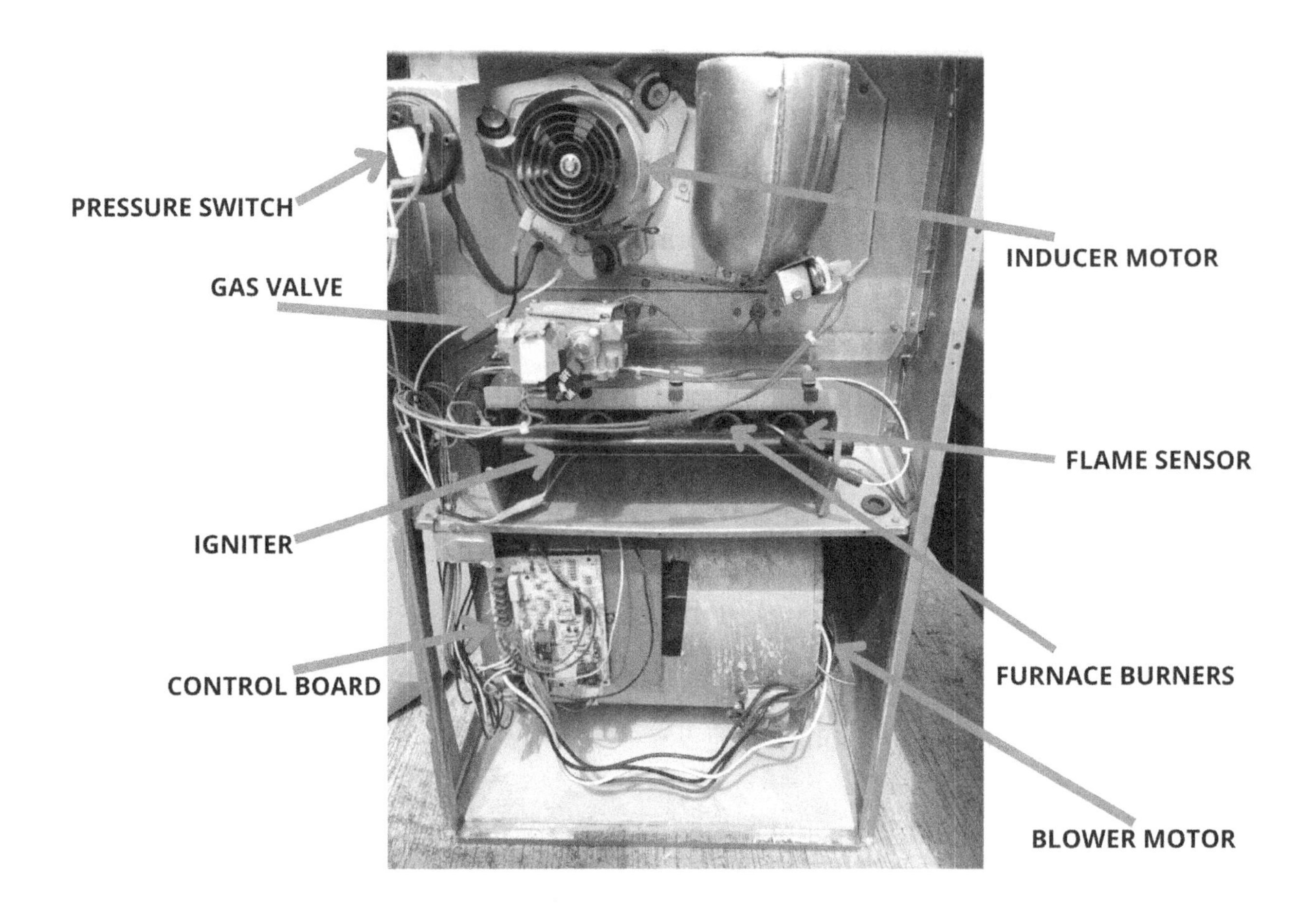

First what I want you to recommend is to read the furnace manual. If you don't have it, then you can go online and find it for your specific furnace model. What you should look for is **the sequence of operation**. It will help you to understand how your furnace works in detail, identify all components, and help you to troubleshoot the system.

Most of them provide recommendations on what the furnace will do when something goes wrong in the system. Down below I provide some basic troubleshooting process for beginners.

But first, let's look at the basic sequence of operations for a gas furnace. The sequence for your furnace model may be slightly different.

Sequence of operation for a direct ignition gas furnace:

1. Thermostat goes into a heating mode and send signal to the control board. On the control board "R" wire starts connecting to "W" wire and sends 24 V to start the sequence of operation.
2. Inducer motor turns on. An inducer motor is essentially an additional fan that blows air into the furnace to ensure efficient combustion of gas in the furnace.
3. Pressure switch proves that inducer is running and that there are no blockages into exhaust.
4. The igniter turns red and starts ignition (or an electric spark may occur with another type of ignition).
5. The gas valve starts to operate, allowing gas to enter the furnace, and the gas starts to burn.

6. Starts the flame identification process to verify that there is a flame and not just gas coming out (flame sensor or flame rod is used for this).
7. The fan motor starts blowing air through the heated heat exchanger. Typically, there is a delay in starting the blower motor to allow the heat exchanger to warm up. This delay is set on the control board.

A heat exchanger is a series of metal tubes that are heated by the combusted gas and have high thermal conductivity (the ability to transfer heat). Modern furnaces also have a so-called secondary heat exchanger, which increases the efficiency of the furnace to 90-95%. The secondary heat exchanger is located after the first one and allows to take away the remaining heat from the gas that was not absorbed by the first one, as well as water vapor in the combusted gas begins to turn into water (condense) and release additional heat. That is why these furnaces are often called condensing furnaces.

Always check if the furnace is giving any errors (write down what you see – number of flashes in the furnace) - check the diagnostic chart on the back of the furnace doors or inside on the control board. It must be out there somewhere.

Whenever you do anything to the electrical components, turn off the power and check that there is no voltage in the wires.

The Most Common Problems and Repairs that Might Occur

Turn the furnace on and off to check that the furnace doesn't lock up for 3 hours (this often happens when a component fails). But before you do this, check for errors (count the number of flashes) if it is shown on the furnace and write it down so that you could identify a problem number.

Most furnaces have a safety switch, when you remove the furnace door, the power to the furnace is turned off. In order to see how the furnace works with the door removed, we need to hold this switch with a magnet or tape so that we could see the sequence of operation and at what stage the problem occurs. You can find this switch somewhere on the frame where the door is installed.

But when a problem is detected and you want to do something with the system components (replace them), the power supply to the furnace must be turned off.

1. The most common problem that occurs with a furnace is **the dirty flame sensor.** It is probably about 70 to 80% of the time.

In this case, the igniter starts glowing, and the flame occurs, but in a few seconds (mostly from 3 to 10 seconds) the furnace turns off. When the furnace does this several times, a three-hour lockout can occur. The sensor inside does not sense the flame well (the sensor does not heat up to the required temperature), and it seems that there is no flame. A flame sensor is simply a metal rod (it can be bent or not).

All you need to do is to take that sensor out of the furnace and clean it so that it starts shining. To do this, use sandpaper or a dishwashing sponge (stiff side). Usually, the flame sensor is located on the opposite side of the igniter.

If the igniter is on the right side, then the flame sensor is on the left side. The reason why the flame sensor is on the other side is to make sure that all gas burners are lit. The igniter lights the first burner and then all the other burners start to burn one by one, as they have a small channel that allows a small amount of gas to move from one burner to the other until all burners are lit so that the flame sensor signals that everything is okay. If this small channel is blocked by a build-up of combustion residue, not all burners will light, and the furnace will shut down.

In this case, we need to clean the burners and the channel that runs through all the burners. It is usually located at the end of the burner where the flame comes out.

Depending on the size of the furnace, it can have from 2 to 6 burners. Usually, all you need to do is remove the burners from the furnace by unscrewing a few bolts and clean the burner ends and their channels with a wire brush. You can also blow into the burner or use compressed air to remove dust. But some furnaces may not have such easy access, so you need to check how this can be done in your case and read the manual for the furnace.

When the burners are clean, the flame should be predominantly blue. If you see a yellow flame, it means that the burners should be cleaned, and if this does not help, it may mean that there is something wrong with the heat exchanger.

2. High limit switch - prevents the furnace from overheating and catching fire ("high limit is open" – fault may occur). This can be caused by a clogged air filter or a stalled fan motor. A sign of this problem is that the furnace cannot heat the house to the desired temperature due to poor airflow.

The upper limit is made up of two metals, and if the furnace overheats, one of them deforms and this separates them, which disrupts the electrical signal, and the furnace shuts down. It is not difficult to find it, usually it is a round disk, but sometimes it can be a black square. You need to read the furnace manual for your model to know all your components and where they are located.

Often all you need to do is to tap a few times the high limit switch with an insulated screwdriver - this will help to connect those metals together.

3. **Blower motor failed.** The sign of this is a plastic smell in the house, if you noticed it then you need to check the blower motor. If you go to your furnace, you will probably see that everything is working except the blower motor. Check the flashing on the furnace and what it means by reading diagnostic chart. It will help a lot and save you a lot of time.

Your blower motor should start from 40 to 60 seconds after the furnace starts working. In this situation, the furnace starts overheating and the high-limit switch will trip.

If you turn off the power to the furnace and check the blower motor with your hand you will probably feel that it is very hot. If it isn't hot then you need to check if power is sent to the motor with a voltmeter or voltage pen and if on the control board, all fuses are good. If all is well, you can perform a visual check and try replacing the fan motor capacitor. If nothing works replacement of the blower is needed.

4. Inducer motor failed. The control board on the furnace will show the "pressure switch open" code. The furnace in this case doesn't do anything. If you touch the inducer motor by hand and feel that it is hot (the power is sent to the inducer) then this is almost for sure that he is dead, or something is stuck inside. But before doing anything with blower motor make sure that the power is off.

To fix this error, we need to check if there is anything inside and remove it or buy and replace it with a new one.

It is almost always a good idea to check if all wires have a good connection to the terminals (with turned-off power). If there is bad contact, it will cause heating and can catch fire or stop powering the component of the system.

The failures that I described above are probably 90 or 95% of all failures that happen to the furnace - if you can't understand what is going on and find which component doesn't work, then you need to ask some HVAC technician.

But in most cases, you can troubleshoot and repair the system, save hundreds of dollars, and give your home some heat.

MAINTENANCE OF THE HVAC SYSTEM

The purpose of maintenance is to ensure high efficiency and performance, as well as to extend the service life of the system. Maintenance helps to prevent system component failure and keep the system running smoothly. The cost of preventive maintenance is only a small part of the repair.

Filter Replacement: Filters should be checked monthly and replaced as needed, at least every 3 months when you don't use the HVAC system often. Dirty air filters can restrict airflow, which can lead to system failure and poor efficiency. You may notice an increase in energy bills because the indoor air blower will work harder to ensure the required airflow.

Duct and evaporator coil cleaning: Dust and debris in the ductwork can reduce air quality and system efficiency. It is recommended to hire a company to clean your evaporator coil and ductwork every 3 to 5 years.

Outdoor Unit: Make sure that the area around the outdoor unit is free of debris and that nothing obstructs the air flow. Clean the condenser fins and coils regularly to maintain efficiency and prevent overheating or other failures.

Inspect the refrigerant lines: Check the suction line insulation and for leaks in both lines.

Check the drain line: Check that the water drains well and clean the pipes if necessary.

Spare Capacitor: It is recommended to have some spare capacitors for the outdoor unit. This is not maintenance, but it will be very useful to have them during the heating or cooling season. It is an easily replaceable system component that very often fails.

Lubrication: Some systems require periodic lubrication of moving parts of the system. Refer to your HVAC system's operating manual or contact a technician to determine the correct type of lubricant and how often to refill it.

Professional Services: Do-it-yourself maintenance can only go so far. Hiring a licensed HVAC professional to recharge or replace the refrigerant in the system, as well as perform regular maintenance, ensures that the system is inspected, cleaned, and adjusted to manufacturer specifications.

KEY DIFFERENCES BETWEEN RESIDENTIAL AND COMMERCIAL HVAC SYSTEMS

Installing commercial HVAC systems is a complex task that requires multiple specialists such as HVAC engineers, contractors, and architects.

The basic principle is the same: a refrigerant is used to transfer heat. And air or water is used to transport the heat to the outside of the building.

There are some differences:

1. **Scale:** Commercial systems are larger and more complex than residential ones. Covers large spaces such as office buildings, shopping centers, or industrial buildings.
2. **Type of Systems:** While commercial systems are often packaged and located on rooftops, residential systems can be split.
3. **Energy Consumption:** Commercial systems have higher energy demands.
4. **Customization:** Commercial systems allow more customization to fit the specific needs of a building.
5. **Maintenance Requirements:** Commercial maintenance is more specialized and frequent.

In commercial buildings, different areas may require individual temperature, humidity, and air quality control. In a large office building, different temperature settings may be required for different floors or rooms. Residential building usually has uniform requirements for all rooms.

Commercial buildings often use larger and more complex equipment, which often includes cooling towers, boilers, and specialized air handling units. Residential buildings, on the other hand, have simpler split systems or heat pumps. Also, in many cases, commercial buildings require integration with other systems in the building like fire control, building management system (BMS), and security.

The HVAC systems in commercial buildings should be compliant with local building codes and energy efficiency standards.

SUMMARY

Nowadays, fossil fuel prices are rising around the globe and governments are looking for alternative energy sources. However, there is a global trend towards renewable energy, so more and more people want to install heat pumps because they are very efficient in cooling and heating homes, the government offsets part of the cost of installation, significantly reduces energy bills, and reduces the carbon footprint.

The HVAC system must be properly designed and installed, and therefore it is not recommended, and in a lot of cases not legally to install a heat pump system yourself, always check the legal aspect of doing it by yourself.

Heat pumps are more expensive because of the difficulty of installation, complex technology, and design. One of the reasons for its high cost is that many components of heat pumps are located outside the house.

It is unlikely that we will see a dramatic breakthrough in HVAC technology any soon, but only modest, incremental improvements.

With the growing awareness of climate change and the need to decrease dependence on energy resources, there is no doubt that heat pumps are on the rise and becoming more accessible and affordable.

Troubleshooting and repair of an HVAC system is something that you as a homeowner/beginner in 70 to 80% of cases can do yourself. The most common repairs are replacing the capacitor, loose electric wires, blown fuses, and a clogged condensate drainpipe.

All these problems can be solved by a homeowner who understands the basics of HVAC systems. Hopefully, after reading this book, you will be more confident in handling your HVAC system.

Maintenance of the HVAC system is also not something complicated and should be carried out by the owner regularly to prevent unexpected breakdowns or a drop in the efficiency of cooling or heating the living environment.

This beginner's guide is not something that will make you a professional and replace years of studying HVAC systems, training, and hundreds of hours of practice in the real world.

We must have the right expectations from reading one book.

Hope you have gotten some value here.

Good luck and quick fixes in your HVAC system!

Best regards!

To receive a PDF bonus (for use on your phone or laptop) with repair instructions and images from the book, please send me an email with your book purchase confirmation:

goreadandmakeit@gmail.com

Made in the USA
Las Vegas, NV
18 October 2023

79137226R00063